D'oh! Fourier

Theory, Applications, and Derivatives

Primers in Electronics and Computer Science

Print ISSN: 2516-6239
Online ISSN: 2516-6247

(*formerly known as ICP Primers in Electronics and Computer Science* — ISSN: 2054-4537)

Series Editor: Mark Nixon *(University of Southampton, UK)*

This series fills a gap in the market for concise student-friendly guides to the essentials of electronics and computer science. Each book will address a core elements of first year BEng and MEng courses and will stand out amongst the existing lengthy, dense and expensive US textbooks. The texts will cover the main elements of each topic, with clear illustrations and up-to-date references for further reading.

Published

Vol. 5   *D'oh! Fourier: Theory, Applications, and Derivatives*
          by Mark Nixon

Vol. 4   *An Elementary Approach to Design and Analysis of Algorithms*
          by Lekh Raj Vermani and Shalini Vermani

Vol. 3   *Image Processing and Analysis: A Primer*
          by Georgy Gimel'farb and Patrice Delmas

Vol. 2   *Programming: A Primer – Coding for Beginners*
          by Tom Bell

Vol. 1   *Digital Electronics: A Primer – Introductory Logic Circuit Design*
          by Mark Nixon

Primers in Electronics and Computer Science Vol. 5

D'oh! Fourier

Theory, Applications, and Derivatives

Mark Nixon
University of Southampton, UK

World Scientific

NEW JERSEY · LONDON · SINGAPORE · BEIJING · SHANGHAI · HONG KONG · TAIPEI · CHENNAI · TOKYO

Published by

World Scientific Publishing Europe Ltd.

57 Shelton Street, Covent Garden, London WC2H 9HE

Head office: 5 Toh Tuck Link, Singapore 596224

USA office: 27 Warren Street, Suite 401-402, Hackensack, NJ 07601

Library of Congress Cataloging-in-Publication Data
Names: Nixon, Mark S., author.
Title: D'oh! fourier : theory, applications, and derivatives / Mark Nixon,
    University of Southampton, UK.
Description: New Jersey : World Scientific, [2022] | Series: Primers in electronics and
    computer science, 2516-6239 ; vol. 5 | Includes bibliographical references and index.
Identifiers: LCCN 2021048499 | ISBN 9781800611108 (hardcover) |
    ISBN 9781800611191 (paperback) | ISBN 9781800611115 (ebook for institutions) |
    ISBN 9781800611122 (ebook for individuals)
Subjects: LCSH: Fourier analysis. | Fourier transformations.
Classification: LCC QA403.5 .N59 2022 | DDC 515/.723--dc23/eng/20211108
LC record available at https://lccn.loc.gov/2021048499

British Library Cataloguing-in-Publication Data
A catalogue record for this book is available from the British Library.

For any available supplementary material, please visit
https://www.worldscientific.com/worldscibooks/10.1142/Q0328#t=suppl

Typeset by Diacritech Technologies Pvt. Ltd.
Chennai - 600106, India

Much of this book was written during the coronavirus pandemic and it reinforced what a great street we live on in Highfield. This was especially so in these difficult times, and the morale has been great. For the excellent spirits, great help and convivial bonhomie we enjoy here: guys and gals, this one is for you.

Contents

Preface

STYLE

This book is aimed to help improve understanding of the Fourier transform (FT): it is a guided tour, with applications, pictures and some jokes. There are three elements to this material:

i) the book itself;
ii) its website https://www.southampton.ac.uk/~msn/Doh_Fourier/ and
iii) some Matlab software.

The maths is aimed to present the material in the simplest manner. The website contains all the software, from which most of the diagrams were derived, and some other material we hope you find useful. The software has been written in Matlab so you can see working equations and it also makes the book reproducible (via open-source software), which is a welcome theme in modern research. There is colour throughout because it is targeted at electronic reading rather than print copy. An added advantage of electronic reading is its help with any obscure words. Some of this material is repeated in the obligatory 'Why was this book written' Section in the Introduction, because if it was all here, some might miss it. The book is certainly a bit of a romp through some pretty complex material and the style aims to make it digestible. Education should be challenging and fun. If you don't like the jokes, it's just tough &#%@!.

TARGET AUDIENCE

The level here is aimed at undergraduates with a mathematical background who cover Fourier as part of their undergraduate curriculum. As the book covers 1-D and 2-D signal analysis, the target curricula include courses on signal processing and on communications, speech analysis and understanding, and image processing and computer vision, amongst others. Fourier has a wide set of applications indeed, as we shall find. The book is also aimed at people who are interested in furthering their knowledge, for whom Maths might be less practised, and so we include diversions around some of the maths bits (and yes, others might use them too). So there is maths in this book.

That's what will put our astronauts on Mars and food on our plates, so don't knock it. This book is a guide rather than a prescription and the aim here is to be as succinct as possible, with plenty of diagrams.

OVERVIEW OF STRUCTURE

The structure here is essentially lexical, starting with the basics and ending with some more advanced stuff and applications. The structure and background material is exposed in Chapter 1. This also describes the basic nature of the FT and some of the applications, which concurrently describes the book's structure. You should be able to start the book anywhere: it could also be read back to front (intentionally, not just for a different market), so let us pick out some possible starting points. One could start with the applications (which contain less maths) covered in Chapter 6. There are more maths when we start with the continuous FT in Chapter 2. Given that computers are ubiquitous, some might prefer to start with sampled data, which is covered in Chapter 3. If you prefer to see material covered as images, then start with Chapter 4. It is unlikely that any would prefer to start with variants and derivatives of the FT, covered in Chapter 5, though these are often used in applications, as described in Chapter 6. Chapter 5 also includes wavelets and they could subsume the more traditional windowing functions of earlier chapters. Chapter 7 describes Fourier and his colourful life, together with some of the context in which his transform was developed. It gives historians a place to start too. For ready reference, there are Chapter 8, which tabulate basic material. The References are collocated at the end of the book, prior to its Index.

> I was once at a (University) cocktail party where an old dragon asked me what I did so I replied that I am a research mathematician (OK, a bit of a stretch – but not an especially big one). "Oh my" replied the dragon "I always thought maths was a waste of time". Well, I didn't especially like her (or the way she avoided challenge), so I replied "if I said 'I couldn't be bothered to learn how to read or write', you'd think I was really thick". "Ooh rather" replied the absence of neurons, and then she realised what I'd said. "Ooh I say, that's rather rude" was her parting sally. That was one Christmas card less.

IN GRATITUDE

Naturally, I must thank many people. I remain grateful to the many students who have survived my teaching in the past 30 odd years on BSc, BEng, MEng and MSc courses within the Department of Electronics and Computer Science at the University of Southampton, and to my many PhD students. Those who have read all, or parts, of it include

Dr. John Carter, Professor Kevin Bowyer, Dr. David Heathfield, Dr. Seth Nixon, Dr. Alberto Aguado, Professor Rob Maunder, Dr. Vishal Patel, Dr. Sasan Mahmoodi and Michael Beale, my WSP Editor, Ramya Vengaiyan and his team, and the reviewers. Your advice was superb, and your comments were most welcome and have helped to enrich this text. It is a much better book with your help: many thanks. Naturally, living with an academic can be a bit of a pain, so I remain grateful to the forbearance, fine food and excellent company of my wife Caz, and the 'kids' Seth and Nimi.

Writing a book takes a long time and a lot of ideas, some of which might not originally be one's own. If any idea here arose from an unattributed conversation or one of the many web searches made during its writing, obviously it has been forgotten but please accept some apologies as it would have been attributed if possible. Naturally, errors will appear in the final version by virtue of the production though some are invariably human. For this, apologies again and an up-to-date list will be included on the book's website https://www.southampton.ac.uk/~msn/Doh_Fourier/. If you're the first to find them, you'll be sent a pint of good English real ale for free. That's a promise in all of my books, as it actually helps to debug the book. Fourier brings a new dimension to understanding, as you'll find herein. And to you, dear reader, thanks for picking it up. Enjoy!!

Mark Nixon

Southampton, UK
March 2021

Key points (tldr)

Rather than bury some of the most important points within the text, they are listed separately here and repeated later at the appropriate point. There are other points, and some of the complexities are not alluded to here (we have stated the key points as succinctly as possible, using italics for emphasis). If you are new to the subject, move straight-away to page 1 (and come back later). If you are reasonably confident, after reading this move to page 11; if you want to add another point or have a quibble with their semantics, go to Chapter 7.

# (Location)	Key point
1 (Section 1.6)	A *signal* is constructed by addition in the time domain and its *Fourier transform* (*FT*) is determined by separation in the frequency domain.
2 (Section 2.1.5)	A *transform pair* means that a signal that exists as signal A in the time domain with a transform signal B in the frequency domain also exists as signal B in the time domain and signal A in the frequency domain.
3 (Section 2.3.1)	The dual of *convolution* in the time domain is multiplication in the frequency domain, and vice versa.
4 (Section 2.5.3)	For *windowing* a periodic signal, choose a window length of three times its period.
5 (Section 3.1.2)	*Sampling*: in order to be able to reconstruct a (discrete) signal from its samples we must sample at *minimum* at twice the maximum frequency in the original signal.
6 (Section 3.2.3)	*Replication*: the transform pair of the sampling function implies that the spectrum of a sampled signal repeats indefinitely in the frequency domain.
7 (Section 3.6.4.1)	The *fast Fourier transform* (FFT) gives the same result as the discrete FT and for N points it is $\frac{N}{\log_2 N}$ times faster (and when N is large that means a lot faster).
8 (Section 5.1)	*Variants of the FT*: the FT is based on sine and cosine waves and uses only one of an infinite number of possible basis function sets.

(Continued)

(*Continued*)

# (Location)	Key point
9 (Section 5.7.2)	*Wavelets* allow sensitivity to frequency/time and frequency/space and allow for multi-resolution time-scale analysis.
10 (Section 7.1.1)	There is no universally accepted unique *definition* of the FT.

MTFBWY (where the F is Fourier). Lock 'n load, let's roll.

CHAPTER 1

Basic Notions and the Nature of the Fourier Transform

1.1 WHY READ THIS BOOK?

Why on Earth read – or write – another book on Fourier? There are already many books on the Fourier transform, and they are well clever and sometimes superbly illustrated. Well, it is time for a new style, as you will find here. In the usual presentation, the Fourier transform is complex stuff (as we shall find, that is the first in a series of terrible puns) and is often met first at the university level in courses in Mathematics or Applied Sciences. Fourier tends to come rather unexpectedly because it is very different from other material (a lot of university courses largely contain advanced extensions of material that has previously been encountered at high school). As Fourier is complex and entails quite a bit of maths, some students find it opaque and rather indigestible (sometimes the voluminous maths makes it rather dry too). But, as we shall find, it is left field and allows you to think of things in a completely different way, and that is good. That is what education is for: to introduce new insights and ways of thinking. Fourier happily straddles Engineering and Science, from Chemistry to Mechanical Engineering. Its properties also allow for many different operations, giving capability for use in many different applications. By the end of this book, you'll be assured that it is very good stuff indeed.

This book is aimed at students as well as at anyone with a bit of maths cluttering their mind. Education often appears as an exclusive process, which it simply should not be. Education should be inclusive, for everyone, and it doesn't stop either. The book is structured so that the broad framework is covered before the detail. Education should be fun too and related to the real world so we use analogies wherever we can. This might be the only book on Fourier that includes the human ear, which is a pity because it has an intimate link with transforms.

We also include examples from human and computer vision, not just because they are from Mark's main research area but because they extend something that can sometimes be buried in maths to something that is very much already in our real world. (The phrase 'real world' always seems rather tautological – it must be real since it exists.) There are plenty of applications described here too.

We present a compromise between theory, implementation and practice, covering basic tutorial material and advanced contemporaneous material. As we readily admit, and as you will find, there are maths in this book and they tend to stymie a light-hearted style. We shall do our best to entertain as well as inform you; we assure you that none of the maths is gratuitous. To attempt to help those with fewer maths (say from the first-year University Science, Applied Science or Mathematics itself) or those who tire of it, there are diversions around the nasty bits denoted by footprints (and an example comes very soon).

As an example of how juvenile the humour can get, note that the symbol ⊛ looks like convolution that we'll meet later on – but it's not: it does not exist in maths and it's just a picture of Mark's butt! That's the lowest example of any humour here, and thanks to Kurt Vonnegut [1973] who first used it that way.

There is naturally much material on the web. When we use it directly we provide a reference to that work. Sometimes the material on the web shares our own educational agenda, and though we are aware of it, we do not cite it directly. There is an excellent Wikipedia entry, but this is more for reference than for educational purposes. For example, Wikipedia's Fourier transform (https://en.wikipedia.org/wiki/Fourier_transform), $\mathcal{F}$, of a Gaussian function is given below in its – in their words – 'unitary version' with 'conventional and angular frequency, ξ and ω', and 'non-unitary angular frequency υ.' (How's that for an opening blast? We bet their natty squiggle looks great on a blackboard, eh.) These are given in comparison with our own simpler (we hope you'll find it so!!) version, which is in terms of angular frequency – and we'll find it in Chapter 2. So the different forms are

$$
\begin{array}{cc}
\text{Wikipedia} & \text{Section 2.1.5} \\
\mathcal{F}\left(e^{-\alpha t^2}\right) = \sqrt{\dfrac{\pi}{\alpha}}e^{\frac{-\pi\xi^2}{\alpha}} = \dfrac{1}{\sqrt{2\alpha}}e^{\frac{-\omega^2}{4\alpha}} = \sqrt{\dfrac{\pi}{\alpha}}e^{\frac{-\upsilon^2}{4\alpha}} & \mathcal{F}\left(e^{\frac{-t^2}{2\sigma^2}}\right) = e^{\frac{-\sigma^2\omega^2}{2}}
\end{array}
$$

$$(1.1)$$

OK Wikipedia has more details and so do we later, but this is enough to show that ours is (aimed to be) much simpler; the objectives of Wikipedia differ from our own because it has a factual agenda and ours is educational. We aim to introduce the Fourier transform and then to apply and build on it, which removes some of the early clutter. We seek to explain and develop the material and describe it in digestible form: Wikipedia provides excellent and detailed reference material **after** you have read this book! We do not shirk from maths, as no one should, though we aim to keep it simple and consistent. (The old joke is that there are three sorts of people in the world: those who can do maths, and those who can't.) We give derivations and demonstrations, not proofs; the concentration here is on application and visualisation rather than on complete rigour. Alternatively, if the left-hand side of Equation 1.1 appears less complex than the right-hand side, this book is not for you. You now have an elegant doorstop instead, so: bye-bye, and use it to prop the door open.

In any book, there is also a question of presentation and order. We have here retained a lexical order (it is a book after all) and used forward/backward referencing to retain linkage of topics that cross chapter boundaries. Colour has become more important in textbooks now that colour printing is cheaper, and because electronic versions proliferate, colour is used wherever possible – whilst aiming to satisfy monochrome viewing/printing too. The references are before the Index at the back, though the electronic versions are naturally searchable. One of the references is the one that started it all off [Fourier, 1878]. And if you think the spelling's duff, we have checked it like mad though, with apologies, some typos might remain. It is worth emphasising that it is written in English (with a guide to the informal English on the book's website https://www.southampton.ac.uk/~msn/Doh_Fourier/): don't let the spelling colour your view.

1.2 SOFTWARE AND REPRODUCIBILITY

The focus here is very much applied: equations on their own are of little use, and more is learned by applying the maths. In this respect, we also provide code that you can use to apply and extend your knowledge. We can assure you that sometimes it is very difficult to write code from a presented set of equations, and some of that frustration is repeated in the code herein. Aiming to avoid that, all functional equations here have been implemented. The code is aimed to be *reproducible* so you can generate as much of the results as can be done (some of the earlier works are harder to reproduce, especially the material that uses continuous maths).

The software here is written in *Matlab*. This is widely used in universities and there is a public domain version called *Octave* (because

Matlab is bloody expensive!). The software code is written for educational purposes and is available from this book's website https://
www.southampton.ac.uk/~msn/Doh_Fourier/. The purpose here is
illustrative for educational purposes; if you use it for anything else, you
are most welcome, but note that we admit no liability. The code uses
hardly any of Matlab's packages and uses basic functionality only. Each
piece of code is intentionally separate, and there is no fancy GUI. It is
simply there, so you can get the same results and play with parameters to see what they do. Reproducibility, which implies code and data
availability, is a strong and most welcome thrust of modern research.
This book is not about the foibles inherent in any computer package
as there invariably are some: any ramifications of the Matlab/Octave
implementations are described in the worksheets that support each
chapter, and sometimes using Anglo-Saxon terminology.

A selection of Fourier transform software is given in Table 1.1
though there are more than these. There are other computational
platforms such as Maple and Mathematica, and these offer their own
computational resource for Fourier analysis, as does Matlab, though it
is not used here. There are packages of Fourier algorithms in a selection of languages. Note that these are often open source and their
quality often depends on a user community; computational architectures continue to evolve, leading to maintenance issues. Some of these
are aimed purely at speed, such as Fastest FT in the East FFTE and in
the West FFTW, whereas others are more general.

TABLE 1.1 Selection of Fourier transform software.

Vendor	Site	Platform
Wolfram	https://reference.wolfram.com/language/guide/FourierAnalysis.html	Mathematica
Math-works	https://uk.mathworks.com/help/matlab/fourier-analysis-and-filtering.html	Matlab
Maplesoft	https://www.maplesoft.com/support/help/maple/view.aspx?path=examples%2Ffourier	Maple

Package	Site	Language/remit
Scipy	https://docs.scipy.org/doc/scipy/reference/tutorial/fft.html	Python/discrete
FFTW	http://www.fftw.org/	C (and Fortran)/speed
FXT	https://www.jjj.de/fft/fftpage.html	Many/algorithms
FFTE	http://www.ffte.jp/	Fortran/speed

1.3 NOTATION

We shall start being clear on *notation*. Any time a new (technical) term occurs, it will be in italics and put in the Index. We have the following set of symbols summarised in Table 1.2. *Variables* can be numbers or scalars, denoted p. Continuous variables represent signals like speech or the surface of the sea, measured as a continuously varying electronic signal. *Sampled* signals are those stored at discrete times (which are particular time instants) and are formed by sampling the continuous (analogue) signals using an analogue-to-digital converter and then storing them in a computer system. A variable that is a function of continuous time t will be denoted $p(t)$, and when the signal is sampled it is a *vector* (**p**) or a *matrix* (an image) (**P**) of the sampled points. One-Dimensional (1-D) points are usually stored in a vector, which for N elements is denoted $\mathbf{p} = [p_0, p_1, \ldots, p_{N-1}]$. (Note that there are also N points in the interval $[p_1, p_2, \ldots, p_N]$ and we will start from 0 to make the maths simpler, but the Matlab code goes from 1 to N, which will plague us later.) Computer *images* are generally a 2-D matrix of discrete points called *pixels* (picture elements). A 2-D matrix has M rows each of N columns, so an $M \times N$ matrix **P** is

$$\mathbf{P} = \begin{bmatrix} P_{0,0} & P_{0,1} & \cdots & P_{0,N-1} \\ P_{1,0} & \ddots & & \vdots \\ \vdots & & & \\ P_{M-1,0} & \cdots & & P_{M-1,N-1} \end{bmatrix}$$

and in the maths, each element is $P_{x,y}$ (the addressing is 'in the door and up the stairs', as in a map) which is addressed in Matlab as `P(y,x)` (where the swap of `y` and `x` is for efficient storage).

   We will have functions of discrete and continuous variables. If a function is of time t and there are parameters a and b that change the overall function, then to separate the independent variable time t from the parameters, the function will be given as

$$f(t; a, b)$$

Given that the book is about Fourier, the Fourier transform process is denoted $\mathcal{F}()$ and the transformed variable is preceded with F when it is continuous or an **F** when discrete. For continuous signals, the Fourier transform takes us from time t to frequency f, and similarly for sampled signals. There will be more about frequency soon!

   We shall also be using complex numbers throughout (that was the pun at the start). We shall use $j = \sqrt{-1}$ though others prefer i (which is used as the symbol for current in electronic circuits). This might be your first experience of complex numbers and some wonder whether they can even exist because square roots imply positive numbers. Well,

TABLE 1.2 Symbols.

	1-Dimensional	**2-Dimensional**
Continuous variables	$p(t)$	$P(x, y)$
Discrete (sampled)	$\mathbf{p}$ (with elements p_x)	$\mathbf{P}$ (with elements $P_{x,y}$)
Fourier transform	$Fp(f) = \mathcal{F}(p(t))$	$FP(u, v) = \mathcal{F}(P(x, y))$
	$\mathbf{Fp} = \mathcal{F}(\mathbf{p})$	$\mathbf{FP} = \mathcal{F}(\mathbf{P})$

FIGURE 1.1 Complex vector

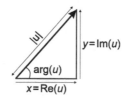

they certainly do exist, because $j^2 = -1$. Complex numbers are often written in vector form

$$u = x + jy = \text{real}(u) + j \times \text{imaginary}(u)$$

where $x = \text{real}(u) = \text{Re}(u)$ and $y = \text{imaginary}(u) = \text{Im}(u)$ are the two components. The vector form is shown in Figure 1.1 for which the length of the vector is its *magnitude*

$$|u| = \sqrt{\text{Re}(u)^2 + \text{Im}(u)^2}$$

and the *phase* is its direction

$$\arg(u) = \tan^{-1}(\text{Im}(u)/\text{Re}(u))$$

where $\tan^{-1}$ is sometimes written (elsewhere) as $\tan^{-1}() = \text{atan}()$. *Allons-y*, it's time to use them.

1.4 BASIC FUNCTIONS

As we shall meet them throughout the book, we shall define some of the functions that are used here. We shall start with some continuous *sine* and *cosine* waves. We plot their forms in Figure 1.2, for a time equal to two *periods* where a period is the time of a complete cycle, the interval between two peaks or troughs (and it's not a US period, period). Sine and cosine waves are also described by their *frequency*, which is the amount they change with time. Part of their popularity now is our electronic world, and it now seems natural that these signals exist. Do we meet sine and cosine waves in real life? Naturally, speech is the transmission of longitudinal waves and is composed of sine and cosine waves. Fourier himself noted the diurnal variation 'the succession of day and night' in the Preliminary Discourse in his great work *The Analytical Theory of Heat* [Fourier, 1878] and though his focus was the equations of the movement of heat, he noted that 'The same theorems ... apply directly to certain problems of general insight.'

We shall need a few symbols; the two functions that include time t and frequency f

> The change in the amount of daylight time is also a sinusoidal function and it goes from midwinter to midsummer and back again, with a period of 1 year. The slope of the change is quite slight, approximately 4 min/day in the UK, except at midsummer and midwinter where there is little change in the amount of daylight for some days.

$$s_1\left(t;f\right) = \sin\left(2\pi ft\right) \qquad s_2\left(t;f\right) = \cos\left(2\pi ft\right)$$
$$(1.2)$$

The relationship between the time period T and the frequency f is an inverse one:

$$\text{frequency } f = \frac{1}{\text{period } T} \qquad (1.3)$$

If we bang a drum every ½ second, then we beat it at a frequency of 2 beats per second. Frequency (measured in *Hertz* (Hz)) is the rate of repetition with time, measured in seconds (s); time is the reciprocal of frequency and vice versa (Hertz = 1/seconds; s = 1/Hz). In Figure 1.2, the frequency $f = 1$ Hz and the period, T (the length of the cycle), is 1 second, $T = 1$.

We shall never shirk from challenges, and that is where we find the footprint diversions, as we do here. These are often where the nastier bits are (or the bits where students often have problems, IMHO). You can skip this here, and meet it again in the next chapter (you can't avoid it!). There are two measures of frequency:

1. frequency in Hertz, $f = 1/T$;
2. *angular frequency* ω in radians (the number of cycles per second, $\omega = 2\pi f = 2\pi/T$);
3. Ouch!

FIGURE 1.2 Sine and cosine waves

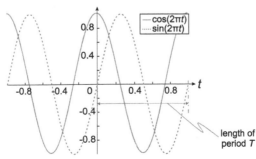

Simplistically, angular frequency will save us writing 2π lots of times. It's more intrinsic than that the period of a sine wave is between where it starts at zero time with $\sin(0) = 0$ and where it ends when $\sin(2\pi) = 0$. Then it starts again, as in the sine wave in Figure 1.2. If we write the function with time as a fraction of the period (i.e t/T), then our sine wave in Equation 1.2 is

$$\sin\left(\frac{2\pi t}{T}\right)$$

So we have a sine wave in terms of frequency and angular frequency as

$$\sin\left(\frac{2\pi t}{T}\right) = \sin 2\pi ft = \sin \omega t$$

Thus ω is a measure of how often a full cycle repeats and its units of radians per second measure the number of full cycles for each second.

One of the fundamental relationships we shall meet a lot is the one between magnitude and phase. The magnitude of a signal describes its size; the phase describes when it occurs. The signals in Equation 1.2 are actually the same but shifted in time by their phase. Denoting the amplitude as A and the phase as ϕ, we have for a signal $s(t)$:

$$s(t) = A\sin(2\pi ft + \phi) \tag{1.4}$$

and for $s_1(t)$ in Figure 1.2: $f = 1$, $A = 1$ and $\phi = 0$; for $s_2(t)$: $f = 1$, $A = 1$ and $\phi = 90°$. The phase is expressed in degrees or in radians ($360° = 2\pi$ radians) and computers invariably use radians (causing quite a few programming whoopsies). The relationship between magnitude and phase is implicit. The magnification of a signal is called *gain* and it cannot exist without phase, and vice versa.

The *Dirac delta* function can be thought of as a function that is infinite at a single point and zero elsewhere. This can be expressed as δ which is a function of time:

$$\delta(t) = \begin{cases} \infty & \text{if } t = 0 \\ 0 & \text{otherwise} \end{cases} \tag{1.5}$$

where the curly bracket denotes an either/or situation. Its representation, Figure 1.3(a), uses an arrow pointing at the sky (to indicate infinity). Its area is unity:

$$\int_{-\infty}^{\infty} \delta(t)\,dt = 1 \tag{1.6}$$

We shall also meet step functions, and the *Heaviside* or *unit step* is a function shown in Figure 1.3(b), that is, 1.0 after a specified time, being zero before that time, so the Heaviside unit step H is

$$H(t) = \begin{cases} 1 & t \geq 0 \\ 0 & \text{otherwise} \end{cases} \tag{1.7}$$

FIGURE 1.3 Basic functions (a) Dirac delta; (b) (Heaviside) unit step;
(c) delayed unit pulse

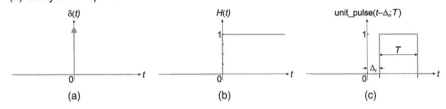

The *unit pulse* is a function that is unity for T seconds and zero elsewhere, giving

$$\text{unit\_pulse}\,(t;\,T) = \begin{cases} 1 & 0 \leq t < T \\ 0 & \text{otherwise} \end{cases} \tag{1.8}$$

And, partly to emphasise notation, a unit pulse of duration T is delayed by Δ_t s. Figure 1.3(c) can be formed from two unit step functions as

$$\text{unit\_pulse}\,(t;\,\Delta_t,\,T) = H\left(t - \Delta_t\right) - H\left(t - T - \Delta_t\right) \tag{1.9}$$

as the pulse is unity from time $t = \Delta_t$ and returns to zero at time $t = T + \Delta_t$ (when the second Heaviside function switches on and is subtracted from it).

1.5 ANALYSING SIGNALS BY THEIR COMPONENTS: APPROXIMATING FUNCTIONS BY MATHEMATICAL SERIES

1.5.1 Taylor series

The Taylor series has more uses than approximation. In Mark's research field of computer vision, the notion that the difference between two points (achieved by subtracting $f(t)$ from both sides of Equation 1.10) in space and in time allows estimation of derivatives (in space – called edge detection, as in Chapter 6).

The concept of splitting a mathematical function into its constituent parts has a long history. It is an attractive concept because it allows us to understand what the function is made up of. We shall start with functions that can be decomposed into their constituent parts (the term 'decompose' might appear unfortunate, but there is no notion of morbidity and subsequent decomposition, it's just a fancy word for splitting). We shall start with the *Taylor series* because it is very useful and serves as an introduction here.

The Taylor series is concerned with representing a function as a sum of parts that comprise its derivatives (rates of change) of different orders. A common use is the estimation of the value of a function f at some future time. The current time is t and we want to predict the value of the function Δt later at time $t + \Delta t$, and the Taylor series is:

$$f(t + \Delta t) = f(t) + f'(t)\,\Delta t + \frac{f''(t)}{2!}(\Delta t)^2 + \frac{f'''(t)}{3!}(\Delta t)^3 + \ldots + \frac{f^n(t)}{n!}(\Delta t)^n$$

(1.10)

where $f'(t)$ is the first derivative and 3! is the factorial $3! = 3 \times 2 \times 1$ (it's not just an exclamation mark, eh!). The value of n defines the number of terms in the series and controls the accuracy of the value computed for $f(t + \Delta t)$. It is an equation and some might be happy with just that; others might prefer it graphically, so Figure 1.4 shows a function and three approximations to it derived from a Taylor series. The first approximation is $f(t + \Delta t) = f(t)$, where we just use the previous value without any gradient information. The approximation becomes closer when we include the slope of the function, the first-order derivative, to obtain $f(t + \Delta t) = f(t) + f'(t) \times \Delta t$. As we include more higher-order derivatives, the approximation becomes increasingly close to the true value.

In this way, we have decomposed a function into its constituent parts. If the summation was to extend to infinity, then the calculated value would be perfect. When we include a reduced number of terms, then the approximation differs from the true value.

Let us see the Taylor series in action; we shall use it to compute a value of the sine function. We need an equation for this so we first consider that $\sin(x) = \sin(0 + x)$ and use the Taylor series to expand the function around zero as

$$\sin(x) = \sin(0 + x)$$

$$= \sin(0) + (\sin(0))'\,x + \frac{(\sin(0))''}{2!}x^2 + \frac{(\sin(0))'''}{3!}x^3 + \ldots + \frac{(\sin(0))^n}{n!}x^n$$

Now, $\sin(0) = 0$ and the differential of sine is cosine so the first-order derivative is $\frac{d\sin(0)}{dx} = (\sin(0))' = \cos(0) = 1$. As the third-order derivative is $(\sin(0))''' = -\cos(0) = -1$, a pattern is set for the odd-order

FIGURE 1.4 Taylor series approximation

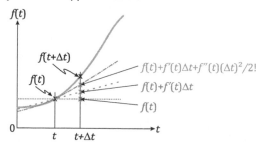

derivatives, alternating between +1 and −1. Looking at the even-order derivatives, $(\sin(0))'' = -\sin(0) = 0$, all the even-order derivatives are zero. Thus, by substituting the derivatives, the equation is

$$\sin(x) = x - \frac{x^3}{3!} + \frac{x^5}{5!} - \frac{x^7}{7!} + \dots$$

We use this equation to compute a value of $\sin(x)$ (where x is in radians) and the accuracy of the computed value depends on the number of terms used. Using $x = 0.9 \times \pi = 2.8274$ radians, we shall seek to estimate $\sin(2.8274)$. Noting that the value of $\sin(2.8274)$ should be small (-ish, because $\sin(\pi) = 0$) and positive, we start at $\sin(2.8274) = 2.8274$ and the first two terms give

$x - x^3/3! = -0.940$ this is miles (or kilometres) out, and now
negative. With the next term

"   " $+x^5/5! = 0.566$ this is much closer (and has the right sign).
Adding the next

"   "   " $-x^7/7! = 0.279$ and the estimate is changing less now

"   "   "   " $+x^9/9! = 0.311$ it's not changing much now

"   "   "   "   " $-x^{11}/11! = 0.309$ and now it's
changing little

In fact, the true value is $\sin(2.8274) = 0.3090487$, so the last value is bang on (to three significant figures). As we include more terms, the computed value becomes more accurate.

Computers could use the Taylor series every time a value of the sine function is needed, but it would be rather slow. Instead, it is more common to use a stored look-up table and interpolate between the stored points as needed.

The Taylor series is a decomposition based on derivatives, and the accuracy depends on the number of terms used. Joseph Fourier had the notion that we could decompose a signal into a series of sine and cosine waves, where the accuracy of the decomposition depends on the numbers of sine and cosine waves used. In his time, this led to the *Fourier series,* which was a solution to the heat flow equation (which was needed so as to better cool down Napoleon's cannons so they could batter the English more effectively) and he was the first to use the word 'transform' to describe the process. Fourier's colourful history is described later in Chapter 7, and we shall use some caricatures derived from his portrait. Fourier might not have been the first to have the notion of a cosine/sine decomposition (perhaps that was Aristotle's first meme), though he was the first to publish it with success.

1.5.2 Fourier series

There are many other decompositions that can be used to represent signals, starting perhaps with Euler. Fourier had the fantastic insight that this could be achieved using sine and cosine waves, and in *The Analytical Theory of Heat* he noted in the concluding sections that 'it remains incontestable that separate functions, or parts of functions, are exactly expressed by trigonometrically convergent series.' As we recognise it now, the generalised form of the Fourier series is made up of summations of cosine and sine waves and is defined as

$$f(t) = \frac{a_0}{2} + \sum_{n=1}^{\infty} a_n \cos(nt) + \sum_{n=1}^{\infty} b_n \sin(nt) \qquad (1.11)$$

where the bias term is

$$a_0 = \frac{1}{\pi} \int_{-\pi}^{\pi} f(t)\, dt \qquad (1.12)$$

and the coefficients are

$$a_n = \frac{1}{\pi} \int_{-\pi}^{\pi} f(t) \cos(nt)\, dt \qquad (1.13)$$

$$b_n = \frac{1}{\pi} \int_{-\pi}^{\pi} f(t) \sin(nt)\, dt \qquad (1.14)$$

By expanding into separate terms we have

$$f(t) = \frac{a_0}{2} + a_1 \cos(t) + a_2 \cos(2t) + \ldots + a_n \cos(nt)$$
$$+ b_1 \sin(t) + b_2 \sin(2t) + \ldots + b_n \sin(nt)$$

We need to find the values of the coefficients a_n and b_n, which scale the sine and cosine waves so as to compose the signal $f(t)$.

To illustrate this composition, we shall calculate the Fourier series of a *square wave* with period 2π and maximum amplitude 1.0 as shown in Figure 1.5. As the amplitude of the square wave is zero for half the time and for the other half it is +1.0, we can evaluate Equation 1.12 in terms of the area of one half as

$$a_0 = \frac{1}{\pi} \times \pi = 1$$

FIGURE 1.5 Square wave

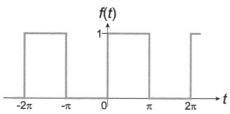

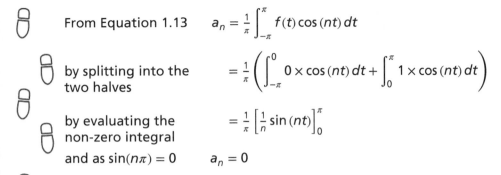

From Equation 1.13 　　　$a_n = \frac{1}{\pi} \int_{-\pi}^{\pi} f(t) \cos(nt)\, dt$

by splitting into the two halves 　　　$= \frac{1}{\pi} \left(\int_{-\pi}^{0} 0 \times \cos(nt)\, dt + \int_{0}^{\pi} 1 \times \cos(nt)\, dt \right)$

by evaluating the non-zero integral 　　　$= \frac{1}{\pi} \left[\frac{1}{n} \sin(nt) \right]_{0}^{\pi}$

and as $\sin(n\pi) = 0$ 　　　$a_n = 0$

So much for a_n. (It always seems like a waste of effort when something complicated evaluates to zero.) Let's try b_n:

by definition, Equation 1.14 　　　$b_n = \frac{1}{\pi} \int_{-\pi}^{\pi} f(t) \sin(nt)\, dt$

by the two halves 　　　$= \frac{1}{\pi} \left(\int_{-\pi}^{0} 0 \times \sin(nt)\, dt + \int_{0}^{\pi} 1 \times \sin(nt)\, dt \right)$

from the non-zero term 　　　$= \frac{1}{\pi} \left[-\frac{1}{n} \cos(nt) \right]_{0}^{\pi}$

Hooray, a result! 　　　$b_n = -\frac{1}{n\pi}(\cos(n\pi) - 1)$

By analysing the result, we can note that

1. If n is even ($n = 2, 4, 6 \ldots$), then $b_n = 0$ because $\cos(n\pi) = 1$
2. If n is odd ($n = 1, 3, 5 \ldots$), then $b_n = \frac{2}{n\pi}$ because $\cos(n\pi) = -1$

This can be summarised as

$$b_n = \begin{cases} 0 & n = 2, 4, 6 \ldots \\ \dfrac{2}{n\pi} & n = 1, 3, 5 \ldots \end{cases}$$

By substituting in Equation 1.11, the Fourier series of the square wave shown in Figure 1.5 is then

$$f(t) = \frac{1}{2} + \frac{2}{\pi} \sin(t) + \frac{2}{3\pi} \sin(3t) + \frac{2}{5\pi} \sin(5t) \ldots$$

We have implemented this equation in Code 1.1 as

$$f(t) = \frac{1}{2} + \sum_{n=1}^{N} \frac{2}{(2n-1)\pi} \sin((2n-1)\,t) \qquad (1.15)$$

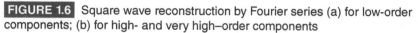

FIGURE 1.6 Square wave reconstruction by Fourier series (a) for low-order components; (b) for high- and very high–order components

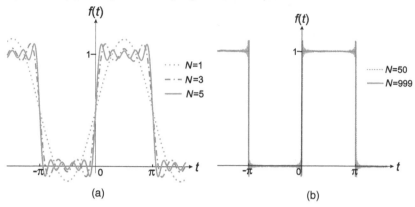

(a)   (b)

and plotted some results for differing values of n in Figure 1.6. Clearly, the low-order components (those with low frequency) in Figure 1.6(a) give the broad structure of the square wave, and the high-order components (those with high frequency) add detail in Figure 1.6(b). If we were to extend this to infinity, we would have a perfect square wave.

There are some odd effects in Figure 1.6(b) that occur when the square wave changes from 1 to 0 (and vice versa) and appears to bounce afterwards. These effects are called *ringing,* which is an oscillation known as the *Gibbs phenomenon,* and occurs because physical signals cannot change from 1 to 0 in zero time (in practice we can't take the Fourier series to infinity, either). The implementation in Matlab providing the results in Figure 1.6 is given in Code 1.1 and uses Matlab's symbolic computation. Note that the ringing continues for many descriptors. If you have the time, try $N = 5999$ and you will find that the ding dongs are still there although they are very small.

Code 1.1 Fourier series of square wave in Matlab

```
syms t %define a symbol for time t
N=3; %let's have descriptors 1 to 5
f=0.5; %start with a0
for n=1:N %add up N descriptors
    f=f+(2/((2*n-1)*pi))*sin((2*n-1)*t); %Eq 1.15
end
fplot(f) %and plot the result
```

So we can indeed make up signals by a combination of sine and cosine waves. This is not yet the Fourier transform because the Fourier series is for analytic functions that are prescribed by equations. Let us now see what the Fourier transform is and how it works. No equations for now as they will come later.

1.6 WHAT IS THE FOURIER TRANSFORM, AND WHAT CAN IT DO?

Essentially, the *Fourier transform* is about finding out how sine waves and cosine waves make up a signal. Given a signal which is a function of time, like the signal $p(t)$ in Figure 1.7(a), there is no way we can determine the frequency content just by looking at it. The signal appears to be repetitive, and it is: this is simply a section of a longer time interval; it is *periodic*, as it repeats. Beyond that, we can suggest it is regular, by internal symmetry, but there is more to be seen. So we use the Fourier transform, which is a set of equations used to process the signal. This gives the amounts of the sine and cosine waves that the signal contains; the Fourier transform can be used to determine the signal's *spectrum*.

The signal $p(t)$ in Figure 1.7(a) was actually constructed by adding three cosine waves of different frequencies, which are those shown (overlaid on the signal $p(t)$) in Figure 1.7(c): $2\cos(t)$, $\cos(2t)$ and $1.5\cos(3t)$. The three frequencies that are determined by applying the Fourier transform are shown in Figure 1.7(b), and they are repeated

FIGURE 1.7 Understanding a signal by the Fourier transform (a) signal $p(t)$; (b) Fourier transform $Fp(f)$; (c) component cosine waves

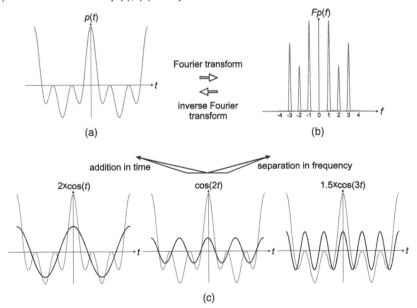

(a)

(b)

(c)

as negative frequency. This is the spectrum. The computed transform is a real one, so as well as the peaks, there are some wobbly bits at the bottom that we shall explain later (that there is positive and negative frequency is a consequence of maths, and in our real world they are the same). As this book is more about getting things working than about theoretical niceties, the transform in Figure 1.7(b) is a real one computed using material found in Chapter 2. The cosine waves provide the signal $p(t)$, a function of time t, when they are added together; the frequencies are separated out by the Fourier transform to derive the transform $Fp(f)$, which is a function of frequency f. Essentially the original signal was constructed by addition in the time domain; the signal is deconstructed by separation in the frequency domain, as in key point 1.

Key point 1

A signal is constructed by addition in the time domain and its Fourier transform (FT) is determined by separation in the frequency domain.

By visual inspection of Figure 1.7(a) we could not determine that the signal was the result of adding three cosine waves and that is the purpose of the Fourier transform. The first frequency in Figure 1.7(c), $2\cos(t)$, describes the overall shape: its peaks and troughs match those of the signal $p(t)$. The higher frequencies describe the data that change more quickly, known as its detail; the higher frequencies appear to match different parts of the signal $p(t)$. In this way, if we want to understand the overall shape of the waveform, we can consider the first component; if we want to consider the detail, then we consider the other components, those of higher frequency. This leads to the possibility of compression and coding (to store or transmit its information in a more compact way) via the transform components as we now have access to different parts of the data.

An important point is that the transform process has not changed the data in any way, only the way it is presented. The original signal can be reconstructed from the transform using the *inverse Fourier transform*. When the inverse transform is constructed from all the frequency components, we return to the original signal. So it is a bidirectional process that does not lose information, and it just changes the representation. The descriptions will start with continuous signals and their Fourier transforms in Chapter 2, for which the maths is (perhaps) more straightforward. As digital implementations are now ubiquitous, we then move on to the computer-based implementations. When the signals are stored in sampled form (in a

There used to be analogue computers and hybrid ones (both analogue and digital) but it's donkeys since that term's been used.

FIGURE 1.8 Transforming an image with the Fourier transform (a) image; (b) Fourier transform

(a)   (b)

computer), we have the *discrete Fourier transform,* which is described in Chapter 3, together with the many aspects of the sampling process (and all the bits that have been left out of the description here).

An *image* is a *two-dimensional,* 2-D, signal that changes along the horizontal and vertical directions, rather than a 1-D signal that changes with, say, time. Colour just makes images here prettier (and adds more dimensions); images we shall process here are greyscale, like Figure 1.8(b). For images, the Fourier transform changes the representation from the spatial domain (the image) into a spatial frequency domain where the components are the frequencies along the horizontal and vertical directions, the amount of change along each axis. The inverse 2-D Fourier transform returns to the original image, with exact reconstruction when all transform components are used. The Fourier transform of the image in Figure 1.8(a) is shown in Figure 1.8(b). As this is our first encounter and it is a point that has perplexed my students in the past, it should be noted that displaying transforms requires more processing and this is covered for 1-D signals in Section 3.2.3 and for images in Section 4.3.1. Figure 1.8(b) looks more like an image of some galaxies, and the structure reflects the structure of the original image. There are also many transform components that have no value, and this is why we can use the transform for coding/compression to reduce the amount of storage (so you don't fill up your cloud space). Images and their Fourier transforms are described in Chapter 4 (together with the complexities of producing – and even viewing – these images, which have not been described here).

1.7   EVERYDAY USE OF THE FOURIER TRANSFORM

We all use transforms all the time, even though we might not recognise this as such. To show this, we shall show how transforms are to be found in speech recognition and speech coding, as well as for images.

It is not just technology and a load of maths: we shall also describe how human hearing transforms sound into signals we interpret as the sound or as the speech. So transforms are indeed part of our everyday life.

1.7.1 Transforms and speech recognition

We shall first consider processing speech as one of the everyday applications of the Fourier transform. For telephone and voice-over IP, it is prudent to code and compress the speech so that the available bandwidth can be better used. If the *speech coding/compression* does not affect the perceived quality, using compression can only improve performance. The transform just shown in Figure 1.7(b) contains only three numbers, whereas the signal from which it was derived in Figure 1.7(a) contains many more (as the signal $p(t)$ is essentially a sequence of numbers). So to transmit the signal we only need to send the three transform components, as long as the receiver has a *decoder/decompressor* that can reconstruct the original signal. This clearly reduces the volume of material that is transmitted.

A speech coder is shown in Figure 1.9. This is part of a standard *coder* (covered in Section 6.7) that first uses the Fourier transform to compress the sound, followed by another transform to compress the information further. This can be used to separate the sound into speaker and into speech, so the same approach is used in speaker recognition (to determine who is speaking) and in speech recognition (to determine what has been said). Speech recognition is used in telephony such as in FaceTime and mobile phones. The second 'frequency transform' is usually a variant of the Fourier transform called the *Discrete Cosine Transform* (*DCT*), which has excellent compression ability, as described in Section 5.1. Speech coding is also used by Alexa and Siri when we communicate with computers, so the Fourier transform is in general use in our everyday life (unless you're a Luddite hermit, but then you wouldn't be picking this book up anyways).

1.7.2 Transforms and image compression

The image transform just shown in Figure 1.8(b) contains considerably fewer numbers than the image from which it was derived, Figure 1.8(a). Images generate enormous volumes of data, and video even

FIGURE 1.9 Transform-based speech coding

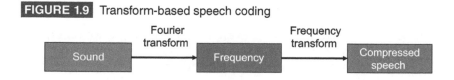

more so. If an image contains $N \times N$ points, that works out as N^2 points. If N is a large number, N^2 is much bigger; a 1000×1000 image in raw form contains 1MB of information. In mobile phones, the available storage would fill up quickly if coding was not used (it fills up too quickly anyways).

A transform-based *image coding* scheme is shown in Figure 1.10, which is part of the *JPEG* system (later described in Section 6.7). In JPEG, the basic transform (a *DCT*) is further compressed according to the level of the components and the information each carries before the coded image is stored. When the image is viewed, a decoder interprets the coded components to provide the images that are presented. There is *lossless* and *lossy* coding, depending on whether none or some of the information is lost, respectively. More space can be saved using the lossy versions, so lossy coding is largely a must for video and films, as a succession of image frames can require very much storage (as Youtube well knows).

The significant components selected from the DCT of the image in Figure 1.8(a) are shown in Figure 1.11(a) and represent 5% of the original transform components. (The transform looks like a quadrant of the transform in Figure 1.8(b): it is described in Section 5.1.) The components are shown as an image in which many points are zero and do not contribute to the reconstruction via the inverse DCT. The image reconstructed from the selected components, Figure 1.11(b), appears to be the same as in Figure 1.8(a) (OK it is greyscale – as will be all the images when considering their transforms – but that is cosmetic only); the difference between the original and the reconstructed images is shown in Figure 1.11(c) and the differences are very small. Because we can return to what appears to be the same image using much less storage, coding is invariably used to store images and video (though it is more sophisticated than the approach here). So we use transforms for images as well as for (speech) signals. Is there anywhere else they are used?

1.7.3 Human hearing and a transform

As Bracewell (one of the doyens of the Fourier transform) once wrote, 'To calculate a transform, just listen' [Bracewell, 1986]. We now

FIGURE 1.10 Transform-based image coding

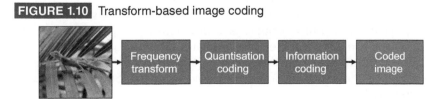

FIGURE 1.11 On transform coding an image (a) 5% of original transform components; (b) image reconstructed from components in (a); (c) image (b) minus Figure 1.8(a)

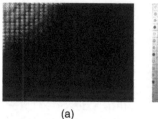 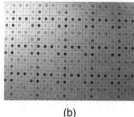 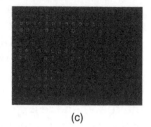

(a)   (b)   (c)

describe in brief how frequency analysis and transforms have an intimate link with human physiology (note that neither we nor Bracewell added 'Fourier' to the word transform, as that cannot be proved). Apologies to any biologists as the coverage here is very brief and basic because this is not a book on biology.

The function of the ear is in essence quite simple: the ear converts sound into nerve impulses that are processed by the brain. The external appearance of the ear is of no interest, we are interested in what happens inside the ear. As shown in Figure 1.12 sound enters through the *outer ear* eventually meeting the *tympanum* (the *ear drum*, the Latin as in the percussion section of an orchestra). The *inner ear* converts the sound to electrical signals. A really neat mechanical arrangement of the *ossicles*, which are the *malleus, incus* and *stapes* (the first two being the hammer and anvil), is activated by the tympanum to stimulate the *cochlea*, where sound is 'heard'. The cochlea consists of channels filled with a liquid and hairs, which are then connected to nerve fibres from the *cochlear nerve*. The sound causes the tympanum to move the ossicles, which pressurises the fluid and causes the hairs to move. Their movement results in *nerve transmissions* to the brain.

> In a shameless plug for our own research, we have recently shown that the appearance of the external ear differs between males and females, though this is unlikely to have any effect on hearing.

Now, if we have flexible rods that are made of the same material and thickness, then it is their length that controls by how much they can flex. This is the case for the hair receptors: they are of differing lengths and respond to different frequencies. Thus, our Fourier series have an anatomical equivalent: our hearing depends on receptors that are tuned to different frequencies. This suggests an interesting perceptual question: if the signals transmitted to the brain are in terms of frequency not in time, then how does the brain decode the information?

FIGURE 1.12 Anatomy of the human ear

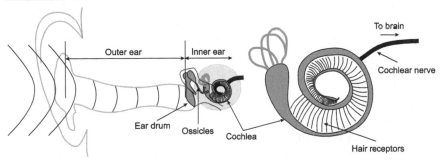

FIGURE 1.13 *The Dark Side of the Moon* album cover

1.7.4 Light and frequency

It is not just our hearing where we experience signals as different frequencies. The cover of Pink Floyd's album (yeah, vinyl, yeah!) *The Dark Side of the Moon* (it's less well known for originally having '*A Piece for Assorted Lunatics*' in its title; it's better known for its music and fantastic sound quality) shows light being split into its component frequencies by the refraction associated with a glass prism (which has been known since Newton's time – or from rainbows). The long wavelengths (red) refract, or bend, the least and the shorter ones (blue) refract the most (Figure 1.13). Later we shall find there is *Fourier spectroscopy*. Transforms exist in light as well as in sound – we meet transforms throughout our daily lives. Optical Fourier transforms are described in Section 6.2.1.

1.8 SUMMARY AND FURTHER READING

This completes our study of the basics. We have introduced many of the underpinning concepts and methods that we shall use throughout the book, such as notation and software. We have also introduced the notions that functions can be approximated by mathematical series and represented by the Fourier series and the Fourier transform. The transform is used to find the sine and cosine waves that make up a signal: we add them up to give a time-domain signal and split them up in the transform. The Fourier transform is used pretty much every day because it is used for image coding and human hearing. We have introduced Matlab for computation and considered the basic signals we shall meet later. There are diversions around some of the maths but never around the software. All that remains for this introduction is the available literature.

We have already quoted Fourier's original book, which has often departed from university libraries now, though a modern presentation is often more incisive (Fourier's book doesn't actually have the Fourier transform, only the series – which is described in Chapter 7). There are many traditional books on these subjects though they lack many tenets of the approach here and are pre-electronic in design. Amongst these, two really stand out: Bracewell's *The Fourier Transform and its Applications* [1986] and *Who is Fourier, a Mathematical Adventure* [Lex, 2012] (references are to be found at the end of the book). They are both simply excellent books, though Bracewell is rather dated now (it goes back to the eighties) and omits a lot of modern material (and especially images and 2-D Fourier transform material) and computers were cumbersome beasts indeed in 1986. *Who is Fourier* is a very enjoyable basic level text, which covers the material well, but at a more basic level (have a look if you have trouble with the maths here). There is an online list of books on the Fourier transform, http://www.ericweisstein .com/encyclopedias/books/FourierTransforms.html, but the texts referenced stop around 2000. That makes them print only, which is a pity as some have some excellent authors. There is the newer *Lectures on the Fourier Transform and Its Applications* [Osgood, 2019] and it certainly has a lively feel and is quite engaging (one of the early subsections is titled '1.4 Two Examples and a Warning') but as its publisher is the American Mathematical Society the mathematical depth might be too high for some: note that the first 100 pages or so are on the Fourier Series. There is also *Discrete Fourier Analysis and Wavelets* [Broughton, 2018] but it is very deep and all discrete, though quite approachable. An excellent and well-established text is Mallat's *A Wavelet Tour of Signal Processing* [2008] ,which has the pedigree of a leading researcher and has a wide coverage (it's nicely peppered with origins of the material too) though it is rather advanced for a novice. Beyond that, a lot of books on the Fourier transform appear rather dated or too general, or delve into heavy-duty signal processing theory. There is an online book of this ilk, which is enormous and very detailed, though published online as a series of web pages, which makes it rather hard to read [Smith, 2011]. For a guide to Matlab, *Essential MATLAB for Engineers and Scientists* [Hahn, 2016] has a long pedigree indeed (though only a short section on Fourier) and there is *Signals, Systems, Transforms, and Digital Signal Processing with MATLAB* [Corinthios, 2009], which has a rather mighty 1300 pages and much of the detail here, but without the style or concision. We must admit bias, naturally, since we have written this book. There is no book like this one, particularly with this style. Let's get cracking with the Fourier transform itself. Read on MacDuff!

CHAPTER 2

The Continuous Fourier Transform

2.1 CONTINUOUS FOURIER TRANSFORM BASIS

2.1.1 Continuous signals and their Fourier transform

The *Fourier transform* (FT) is a way of separating an unknown signal into its component frequencies. It allows us to work out which cosine and sine waves make up a signal; it is similar to the Fourier series, but it is a transform rather than a process. By way of example, our speech is not a single tone but a collection of tones. These are the sine and cosine waves that we can determine using the FT. Consider listening to music: your chosen YouTube delight comes from the computer and is played on speakers after it has been processed via an amplifier. On the mixer settings on the computer, you can change the bass or the treble. Now that is about stored music, which we shall consider in the next chapter. This chapter is about continuous signals, for example, the sound heard from a violin or the output of a microphone. The FT is a way of mapping the signal into the *frequency domain*: a signal varying continuously with time is transformed into its frequency components. When we have transformed the signal, we know which frequencies made up the original sound and if the bass is a bit low, we can spin its dial and boom, boom we go!

> Bass covers the low-frequency components (e.g. from the bass guitar) and treble covers the high-frequency ones (e.g. from the cymbals or from Keith Richard's lead guitar).

We have illustrated this for a signal in Figure 2.1. It is difficult visually to determine the composition of this signal, but it is actually derived from a set of sine waves, and in Matlab, the signal was constructed by `p = a+b*cos(3*t)+c*cos(4*t)`. This is an addition in the time domain. Clearly, the result of summation in Figure 2.1 is periodic, with a hint of sine or cosine waves, but little else can be gleaned directly. The FT shows us precisely what can be gleaned and offers separation in the frequency domain. In the frequency domain, we have a set of

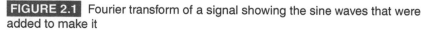

FIGURE 2.1 Fourier transform of a signal showing the sine waves that were added to make it

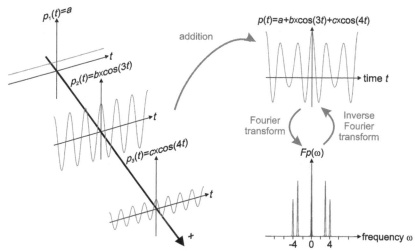

frequencies with different amplitudes (the height of the spikes): the link is the addition of the sine waves at these frequencies and of these amplitudes. In this example, we knew how the original signal was made up. That is the job of the FT: to determine what is inside the signal. There is slight confusion associated with the existence of negative time and frequency. These can be thought of as either mathematical niceties (necessities?) or the time before and the frequency backwards. We could have omitted them for simplicity but have retained them so that the presentation is consistent with later material.

So why do we do this? We have not changed the signal, only its representation. We can now visualise a signal in terms of its frequencies and we can use that knowledge to understand, process or change the signal. The equation that defines the FT, *Fp*, of a signal *p*, is given by a complex integral. We shall build up the equation. First, we have that the FT is a function *a* () of a time-variant signal *p* (*t*)

The FT transform is then	$Fp = a(p(t))$
The FT is a function of frequency so	$Fp(f) = a(p(t))$
$\mathcal{F}$ stands for the FT so	$Fp(f) = \mathcal{F}(p(t)) = a(p(t))$
The function *a* () is actually an integral	$Fp(f) = \mathcal{F}(p(t)) = \int_{-\infty}^{\infty} p(t)\,\text{cms}(t)\,dt$
cms (*t*) describes cosine and sine waves, so	$Fp(f) = \mathcal{F}(p(t)) = \int_{-\infty}^{\infty} p(t)\,e^{-jft}dt$

where cms () stands for 'cos minus sin' because $e^{-jft} = \cos(ft) - j\sin(ft)$ in which j is the complex number $j = \sqrt{-1}$.

OK, that was foot flat down on the accelerator; let us go a bit slower. The FT is a function of frequency and is calculated from a time-domain function. We need to sum up the amount of sine and cosine waves, so there is an integral function. The integral evaluates the amount for all time, so its limits are between plus and minus infinity. The amount is evaluated by multiplying the time-domain signal we are transforming by cosine waves and sine waves. This is expressed in complex form (electronic engineers prefer j to the elsewhere more conventional i – standing for imaginary – because they cannot confuse it with the symbol for electrical current; perhaps they don't want to be mistaken for Mathematicians who use $i = \sqrt{-1}$). This gives an equation for the FT at a particular frequency, and we then repeat this for all frequencies to determine the complete transform of the signal.

Now because we have a complex number, we have a phasor that has horizontal and vertical components. These make up the radius of a circle. As previously described (though perhaps skipped) in Section 1.4, frequency is a measure of how far we go around a circle in unit time so we have *angular frequency* and it is measured in radians per second (2π radians meaning we have gone around the whole unit circle; radians per second measures how many cycles are completed per second). Angular frequency is then denoted by ω where

$$\omega = 2\pi/T = 2\pi f$$

in which T is the period of the cycle. This leads to the conventional definition of the continuous FT

$$Fp(\omega) = \mathcal{F}(p(t)) = \int_{-\infty}^{\infty} p(t)e^{-j\omega t}dt \qquad (2.1)$$

This is the transform illustrated in Figure 2.1. Equation 2.1 is the definition we shall use from now on, though there are other versions, described in Section 7.1.1. As previously described in Equation 1.4, sine and cosine waves are characterised by their amplitude, frequency and phase. A single sine wave does not imply any knowledge of a different sine wave because their characterisations are totally separate and the two signals are said to be *orthogonal*. For vectors this implies that they are mutually perpendicular and changing one cannot change the other. As the set of sine and cosine waves in the Fourier equation are of unit amplitude and are mutually orthogonal, the FT enjoys an *orthonormal* basis set. This is how the transform separates signals in the frequency domain because the frequency components are totally distinct from each other. Listening to different radio stations would be impossible if the separation between frequencies was anything other than complete. As we shall find later there are different transforms

that employ other orthonormal basis sets, providing different properties. For now, we shall stick with the sine and cosine waves of the FT.

Let us see how the transform equation works. We shall start by considering the frequency content of a *pulse* signal. As in Section 1.4 a
pulse is a signal that exists for a short time, equivalent to hitting a desk
(this is very useful in lectures to keep students awake, as those from
the University of Southampton will testify) and it is zero when the desk
is not being thumped. We shall define a pulse $p(t)$ that lasts for T s,
starting at $t = -T/2$ and ending at $t = T/2$. The strength of the pulse
(how hard the desk is hit) is A and it is zero for the rest of time. So

$$p(t) = \begin{cases} A & -\dfrac{T}{2} \leq t \leq \dfrac{T}{2} \\ 0 & \text{otherwise} \end{cases} \tag{2.2}$$

where the curly bracket (curvaceous parenthesis?) denotes an either/or
situation. To obtain the FT, we substitute for $p(t)$ in Equation 2.1. We
choose the limits of the integral as those for which the pulse exists,
from $t = -T/2$ to $t = T/2$, and in this interval $p(t) = A$, so the FT of a
pulse is

$$\mathcal{F}(p(t)) = \int_{-T/2}^{T/2} Ae^{-j\omega t} dt \tag{2.3}$$

The result of integration is $\quad Fp(\omega) = \dfrac{Ae^{-j\omega t}}{-j\omega}\Big|_{t=-T/2}^{t=T/2}$

and by substitution $\quad Fp(\omega) = \dfrac{Ae^{\frac{-j\omega T}{2}} - Ae^{\frac{j\omega T}{2}}}{-j\omega} \tag{2.4}$

By deploying the relationship $\sin(\theta) = \left(e^{j\theta} - e^{-j\theta}\right)/2j$ (a.k.a. Euler's formula) we arrive at the FT of a pulse, which is

$$Fp(\omega) = \begin{cases} \dfrac{2A}{\omega} \sin\left(\dfrac{\omega T}{2}\right) & \omega \neq 0 \\ AT & \omega = 0 \end{cases} \tag{2.5}$$

in which the term for zero frequency (the bias or *d.c.* term for which
$\omega = 0$, as can be seen in Figure 2.1) is derived by the small angle
theorem ($\sin(x) = x$ when x is very small). This function in Equation 2.5
is called a *sinc* function, because

$$\text{sinc}(x) = \frac{\sin(x)}{x}$$

FIGURE 2.2 Fourier transform of a pulse: (a) pulse; (b) sinc

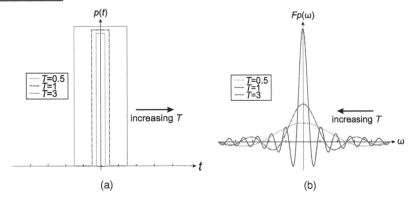

(a)   (b)

so

$$Fp\,(\omega) = AT\text{sinc}\left(\frac{\omega T}{2}\right)$$

A pulse and its transform are shown in Figure 2.2. The pulse is shown with different widths (lasts for different lengths of time) – and is scaled vertically so that the different pulses can be seen. The inverse relationship between time and frequency is shown in the transform: when the pulse is wide the sinc function is thin, and vice versa. When punching the desk, this lecturer points out that when the desk was hit for a long time, the range of frequencies is less than when it was hit for a shorter time. That is what the transforms show: the range of frequencies is less when the pulse is long. Conversely, when the pulse is short there is, proportionally, a greater range of frequencies. As there is a direct link between the two signals via the FT, they are known as a *transform pair*.

2.1.2 Magnitude and phase

As any academic does, we have naturally skipped the truth a bit ("it's obvious", eh!). First off, the transform is a complex number so it has *real*, Re, and *imaginary*, Im, parts

$$Fp\,(\omega) = \text{Re}\,(Fp\,(\omega)) + j\,\text{Im}\,(Fp\,(\omega)) \tag{2.6}$$

The real and imaginary parts cannot be visualised jointly in a 2-D graph and so we have the *magnitude*, denoted $|Fp\,(\omega)|$, and *phase*, denoted $\arg\,(Fp\,(\omega))$, as

$$|Fp\,(\omega)| = \sqrt{(\text{Re}\,(Fp\,(\omega)))^2 + (\text{Im}\,(Fp\,(\omega)))^2} \tag{2.7}$$

$$\arg\,(Fp\,(\omega)) = \text{atan}\frac{\text{Im}\,(Fp\,(\omega))}{\text{Re}\,(Fp\,(\omega))} \tag{2.8}$$

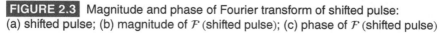

FIGURE 2.3 Magnitude and phase of Fourier transform of shifted pulse: (a) shifted pulse; (b) magnitude of $\mathcal{F}$ (shifted pulse); (c) phase of $\mathcal{F}$ (shifted pulse)

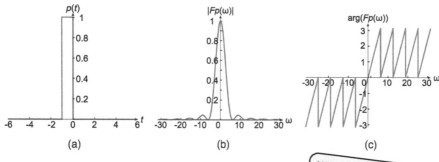

(a) (b) (c)

(Poetic license was not used in Figures 2.1 and 2.2(b) because in those plots $Fp(\omega)$ was purely real, derived from Equation 2.5.) Going back to the lecturer hitting the desk, the magnitude is how hard the desk was hit, and the phase is when it was hit. Not all signals are centred at $t = 0$ (though they often are in books). To illustrate this, and to clarify the magnitude and phase, Figure 2.3(a) shows a pulse that starts at $t = -1\text{s}$ and stops at $t = 0\text{s}$. Its magnitude and phase are shown in Figures 2.3(b) and (c), respectively. Here the magnitude is shown as always positive (Figure 2.2 with the negative bits rectified). The phase increases linearly in the interval $\omega = [-2\pi, 2\pi)$. At $\omega = 2\pi$, the phase appears to return to zero and starts increasing again, but that is simply the wrapping of the arctangent function when the phase returns to zero and starts again.

> We could draw the phase without returning to zero but we'd then have a very big graph, a lot of empty space, and an earful from our publisher

Given that the interest here is explaining the frequency content of signals, we shall leave the phase out for a while, returning later in Section 2.4 where we discuss its importance.

2.1.3 Inverse Fourier transform

What are the amounts of the signals at different frequencies? This is easy for the cosine waves in Figure 2.1 as the amplitude of the cosine waves is related to the height of the scaled delta functions (the spikes) in its transform. But what for the sinc function? We have already met the bias or d.c. term, where $\omega = 0$, so named because the bias is a constant and just shifts the signal up or down. As a signal's frequency increases, it changes more rapidly. By adding together a set of frequency components, we are adding different amounts of change. We shall consider adding up the frequencies that are shown in the transform of a pulse in Equation 2.5. In Figure 2.4(a) the first contribution is the amount of the d.c. component, for which $\omega = 0$. On the

FIGURE 2.4 Reconstructing a pulse from transform components: (a) d.c. component; (b) adding the d.c. component; (c) first frequency; (d) adding the first frequency; (e) second frequency; (f) adding the second frequency; (g) third frequency; (h) adding the third frequency

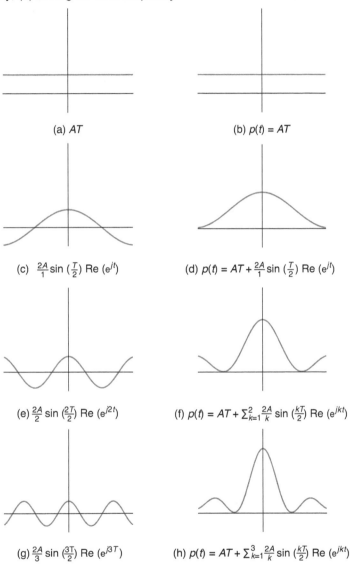

(a) AT

(b) $p(t) = AT$

(c) $\frac{2A}{1} \sin \left(\frac{T}{2}\right) \operatorname{Re} \left(e^{jt}\right)$

(d) $p(t) = AT + \frac{2A}{1} \sin \left(\frac{T}{2}\right) \operatorname{Re} \left(e^{jt}\right)$

(e) $\frac{2A}{2} \sin \left(\frac{2T}{2}\right) \operatorname{Re} \left(e^{j2t}\right)$

(f) $p(t) = AT + \sum_{k=1}^{2} \frac{2A}{k} \sin \left(\frac{kT}{2}\right) \operatorname{Re} \left(e^{jkt}\right)$

(g) $\frac{2A}{3} \sin \left(\frac{3T}{2}\right) \operatorname{Re} \left(e^{j3T}\right)$

(h) $p(t) = AT + \sum_{k=1}^{3} \frac{2A}{k} \sin \left(\frac{kT}{2}\right) \operatorname{Re} \left(e^{jkt}\right)$

left we have the components and on the right the signal made up by adding that and all preceding components, so Figure 2.4(b) is the same as Figure 2.4(a). The first frequency component (Figure 2.4(c)) is a cosine wave of half the period, and when this is added, we start to see the emergence of the pulse (Figure 2.4(d)). We are considering only the real parts of the signal for this reconstruction. The higher frequencies (Figures 2.4(e) and (g)) add increasing levels of detail, closing in on the pulse in Figures 2.4(f) and (h).

This leads to the *inverse FT,* which is used to reconstruct a signal from its transform. In Figure 2.4(h) we have neither the high frequencies nor the negative ones, and we need them to get closer to the pulse. The inverse FT is given by

$$p(t) = \frac{1}{2\pi} \int_{-\infty}^{\infty} Fp(\omega) e^{j\omega t} d\omega \tag{2.9}$$

where the difference from the forward FT is the sign of the exponential function, $e^{j\omega t} = \cos(\omega t) + j \sin(\omega t)$ and the scaling coefficient $1/2\pi$. This integrates all parts of all frequencies to reconstruct the original signal.

The reconstruction of a pulse for $T = 1$ using the inverse FT is shown in Figure 2.5 as the magnitude of the result. When the frequencies are limited to $\omega = \pm 1$ the reconstruction of the pulse is poor, showing what are called *sidelobes* (where the signal appears to bounce, the Gibbs phenomenon earlier encountered with Fourier descriptors) in Figure 2.5(a). When the frequencies are limited to $\omega = \pm 10$ the sidelobes are smaller and the pulse is clearer, though there is still a dip in the middle, Figure 2.5(b). When all frequencies are included, $\omega = \pm \infty$, the pulse is reconstructed exactly, without sidelobes, in Figure 2.5(c).

This is the *transform pair*: there is an intimate relationship between the time and frequency domains and they are invertible. The transform pair of a pulse in the time domain is a sinc function in the frequency domain. Conversely, a sinc in the time domain transforms to a pulse in the frequency domain (or, more practically, a restricted set of data in one domain leads to a sinc function in the other domain). We shall consider other transform pairs later as they show how the transform

FIGURE 2.5 Inverse Fourier transform reconstructing a pulse: (a) $\left|\frac{1}{2\pi} \int_{-1}^{1} Fp(\omega) e^{j\omega t} d\omega\right|$; (b) $\left|\frac{1}{2\pi} \int_{-10}^{10} Fp(\omega) e^{j\omega t} d\omega\right|$; (c) $\left|\frac{1}{2\pi} \int_{-\infty}^{\infty} Fp(\omega) e^{j\omega t} d\omega\right|$

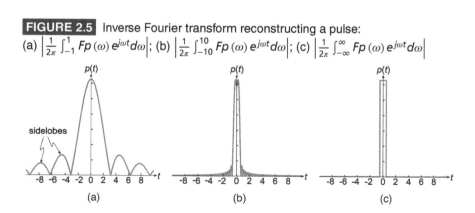

and its maths works. Before that, we must consider implementation in Matlab.

2.1.4 Fourier transform in Matlab

All computation so far can be done with Matlab as shown in Code 2.1. First of all, we shall do symbolic computation and so we define symbols for time `t` and frequency `f`, `w` via the instruction `syms`. Then we specify how long the pulse will last, here 1s (`T=1`). We use the Matlab predefined function `rectangularPulse` to specify the function in Figure 2.2(a), and the continuous FT itself `fourier(p)` for Figure 2.2(b). (Matlab isn't very good at grammar or eponymosity and is case sensitive: `Fourier(p)` fails, `fourier(p)` works!) When the semicolon is omitted at the end of a line, Matlab prints the output, shown here as underlined. We use `fplot` to see the signals, specifying the time (or frequency) range of interest. We generally view the magnitude, as here, and will consider the importance of phase later. Note that the magnitude function is hard-coded (Equation 2.7) because the Matlab `abs` function often introduced discontinuities when operating on symbolic functions (Matlab does an excellent job usually!) When printed, `Fp` is rather complicated and this can be resolved using the `simplify` function which shows `Fp = (2*sin(w/2))/w`, which is what was expected for a pulse of unit amplitude and duration $T = 1$ (Equation 2.5). To evaluate the inverse FT we need to use symbolic integration, `int`, and the limits on frequency need to be specified (the calculation here is by real and imaginary parts, for stability). The complex variable is `1j` (and appears as that when it is copied from Matlab) and if you're happier with `1i` you are welcome to use that. When the limits are infinite, we might as well use Matlab's inverse FT `ifourier`, which takes us back to the pulse we started from. You can use the worksheets to change the functions and see the results, and these are available from this book's website https://www.southampton.ac.uk/~msn/Doh_Fourier/ (as mentioned previously, the worksheets can be used in Matlab and Octave). The intention is that this text is reproducible (all the graphs here are derived from Matlab results and the code which produced them is wholly available) which is a theme of modern research.

Code 2.1 Basic Fourier transforms in Matlab

```
syms t f w %define symbols for time and frequency
T=1; %time duration of the pulse

p = rectangularPulse(-T/2,T/2,t); %define a pulse
fplot(p, [-6,6]) %Figure 2.2(a)
Fp = fourier(p) %let's evaluate the Fourier transform
```

```
Fp = (cos(w/2)*1i + sin(w/2))/w - (cos(w/2)*1i
     - sin(w/2))/w
fplot(Fp, [-10*pi,10*pi]) %Figure 2.2(b)
%Now try a different value for T

%Fp is rather complex. Let us try
simplify(Fp)
ans = (2*sin(w/2))/w
%Ah! That is what we expect

%Inverting the Fourier transform, by parts
q = (1/(2*pi))*(int(real(Fp*exp(1i*w*t)),w,[-10 10])+...
    1j*int(imag(Fp*exp(1i*w*t)),w,[-10 10])); %w=-10 to +10
mag=sqrt(real(q)*real(q)+imag(q)*imag(q));
%we'll plot the magnitude
fplot(mag, [-10 10]) %Figure 2.5(b)

%Finally we go for the inverse Fourier transform
r = ifourier(Fp); %inverse transform the sinc function
fplot(r, [-10 10]) %Figure 2.5(c): we're back where we started
```

2.1.5   Fourier transform pairs

A selection of continuous FT pairs is summarised in Table 8.3, for easy reference. We shall consider some of the basic ones here. The basics of transform pairs are summarised in key point 2.

Key point 2

A transform pair means that a signal which exists as signal A in the time domain with a transform signal B in the frequency domain also exists as signal B in the time domain and signal A in the frequency domain.

2.1.5.1   Delta function

There are some well-known transform pairs beyond the pulse/sinc relationship in Section 2.1.1. We can handle these analytically and they show how the maths works. One assumption is that this leads to the understanding of arbitrary signals, because they can be considered to be a collection of the basic components. First off, as in Equation 1.6 there is a delta function,

$$\delta(t) = \begin{cases} \infty & t = 0 \\ 0 & \text{otherwise} \end{cases} \tag{2.10}$$

so

$$\mathcal{F}(\text{delta}) = \mathcal{F}(\delta(t)) = \int_{-\infty}^{\infty} \delta(t) e^{-j\omega t} dt = 1 \qquad (2.11)$$

As such, an infinitely short pulse leads to an infinite set of frequencies. This can be understood from Figure 2.2 whereas the pulse gets thinner the FT gets wider – take this to infinity (i.e. the pulse is infinitely thin, so the transform is infinitely wide) and you have Equation 2.11.

2.1.5.2   Sine wave

Let us try a sine wave with angular frequency ω_0, $\sin(\omega_0 t) = (e^{j\omega_0 t} - e^{-j\omega_0 t})/2j$

By Equation 2.1 $\mathcal{F}(\text{sine wave}) = \int_{-\infty}^{\infty} \dfrac{(e^{j\omega_0 t} - e^{-j\omega_0 t})}{2j} e^{-j\omega t} dt$

by rearrangement $= \dfrac{1}{2j} \int_{-\infty}^{\infty} (e^{j\omega_0 t} - e^{-j\omega_0 t}) e^{-j\omega t} dt$

and by splitting into two parts $= \dfrac{1}{2j} \int_{-\infty}^{\infty} e^{j\omega_0 t} e^{-j\omega t} dt - \dfrac{1}{2j} \int_{-\infty}^{\infty} e^{-j\omega_0 t} e^{-j\omega t} dt$

$$\mathcal{F}(\sin(\omega_0 t)) = -j\pi\delta(\omega - \omega_0) + j\pi\delta(\omega + \omega_0) \qquad (2.12)$$

because a delta function $\delta(x - \alpha) = \frac{1}{2\pi} \int_{-\infty}^{\infty} e^{jt(x-\alpha)} dt$ (OK, that's a bit of an ouch and we shall get there soon. For now, a delta function has an infinite set of frequency components as in Equation 2.11). The transform of a sine wave is two delta functions (spikes), at the positive and negative frequency of the original sine wave. (We'd be worried if it was anything else: it's the transform of a sine wave.) The transform pairs so far (other than that for the pulse) are shown in Figure 2.6.

2.1.5.3   Gaussian function

The *Gaussian* function defines the *normal* distribution, to honour Gauss' fantastic insight. The Gaussian function/ normal distribution in time is defined in terms of standard deviation σ (or variance σ^2) as

$$g(t; \sigma) = \frac{1}{\sqrt{2\pi\sigma^2}} e^{\frac{-t^2}{2\sigma^2}} \qquad (2.13)$$

The *Central Limit Theorem* suggests that the normal distribution is the result of adding many noisy signals together. (One way to generate random numbers with a Gaussian/normal distribution is to generate numbers with a linear/ flat distribution where all numbers are equiprobable, and to then add them up say 50 at a time. The distribution of the resulting numbers will be normal.) So the normal distribution describes many natural events. One problem is that it extends to

FIGURE 2.6 Fourier transform pairs: (a) sine wave; (b) $|\mathcal{F}$ (sine wave)$|$; (c) delta function; (d) $\mathcal{F}$ (delta)

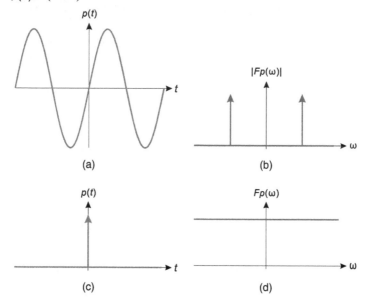

infinity in both directions – more about that later. Here we are interested in the frequency content of the Gaussian function, as exposed by its FT. Let us have a look.

The transform is
$$\mathcal{F} \text{ (Gaussian)} = \int_{-\infty}^{\infty} \frac{1}{\sqrt{2\pi\sigma^2}} e^{\frac{-t^2}{2\sigma^2}} e^{-j\omega t} dt$$

which gives
$$= \frac{1}{\sqrt{2\pi\sigma^2}} \int_{-\infty}^{\infty} e^{\frac{-(t^2+2\sigma^2 j\omega t)}{2\sigma^2}} dt$$

for completing the square
$$= \frac{1}{\sqrt{2\pi\sigma^2}} \int_{-\infty}^{\infty} e^{\frac{-((t^2+2\sigma^2 j\omega t - \sigma^4 \omega^2) + \sigma^4 \omega^2)}{2\sigma^2}} dt$$

by rearrangement
$$= \frac{1}{\sqrt{2\pi\sigma^2}} \left(\int_{-\infty}^{\infty} e^{\frac{-(t+\sigma^2 j\omega)^2}{2\sigma^2}} dt \right) e^{\frac{-\sigma^4 \omega^2}{2\sigma^2}}$$

via the area under a Gaussian curve
$$= \frac{1}{\sqrt{2\pi\sigma^2}} \left(\sqrt{2\pi\sigma^2} \right) e^{\frac{-\sigma^2 \omega^2}{2}}$$

So the FT of a Gaussian is

$$\mathcal{F} \text{ (Gaussian)} = Fg \left(\omega; \sigma_{Fg} \right) = e^{\frac{-\sigma^2 \omega^2}{2}} = e^{\frac{-\omega^2}{2\sigma_{Fg}^2}} \qquad (2.14)$$

FIGURE 2.7 Fourier transform pair of a Gaussian function: (a) Gaussian in time $g\,(t;2)$; (b) $\mathcal{F}$(Gaussian) $Fg\,(\omega;1/2)$

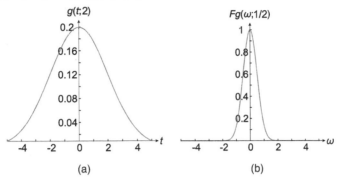

(a)   (b)

The normal distribution can be used to explain many things. For example, people's capability to recognise faces. Many of us are normal and can recognise many people but not all people all the time. In contrast, at one end of the normal distribution there are – few – people with little chance of recognising anyone (called prosopagnosia). At the other end, there are – again few – people who have fantastic recognition and can recognise people even after only a glimpse previously: they are called super-recognisers. In Figure 2.7, if the horizontal axis was chance of recognition (adding and scaling so that the chance varies from 0 to 1) and the vertical axis was the number of people with that chance, then human face recognition ability would perhaps be closer to Figure 2.7(b) than Figure 2.7(a) as the variance would likely be smaller, and there are only few people with prosopagnosia and few super-recognisers.

which is another Gaussian function but with a variance which is the reciprocal of that in the time domain $\sigma_{Fg} = 1/\sigma$. This is expected because time is the reciprocal of frequency. The relationship is shown in Figure 2.7 for $\sigma = 2$. The alternative ways to develop this relationship are to use differentiation or polar variables but these methods are rather more complex. The ouch in this analysis is the integral of the (shifted) Gaussian function, which was the target of completing the square. Analytically, this could be cleaner, but it suffices here. The final relationship is clear, and this can be seen in the Matlab result in Figure 2.7 (note the difference in the scales of the vertical axes, given the scaling factor of the Gaussian function in Equation 2.13).

The Gaussian function has many interesting properties: one is that it is localised both in time and in space. It is not normally possible to localise in time and space at the same time. As we found in Section 2.1.5.1, the FT of an impulse/delta function is a flat function in the frequency domain. That is the extreme case of localisation because it is localised in time but not in frequency. It does actually introduce a further debate concerning the potential relationship between a signal that is contained within a limited set of frequencies (in the frequency domain) and the concentration of a signal in the time domain. This leads to *prolate spheroidal wave functions* which have been suggested to be an improved basis for spectral techniques, but we shall stop here and move on the transform pairs.

2.2   PROPERTIES OF THE CONTINUOUS FT

The properties of the FT are very important in its applications. These properties largely concern factors that affect the appearance or manifestation of signals. If a factor does not change the appearance of the transform of a signal, then the transform is said to be invariant to that factor. Appearance is how a signal is changed; *invariance* is whether that change is manifest in its transform. A selection of the properties of the continuous FT is summarised in Table 8.2, for easy reference.

2.2.1   Superposition

The *principle of superposition* concerns linearity. A *linear* function obeys it. Linearity means that the response of a function to two signals applied independently is the same as that when the two signals are applied jointly. For the FT, this is given by

$$\mathcal{F}(p(t) + q(t)) = \mathcal{F}(p(t)) + \mathcal{F}(q(t)) \tag{2.15}$$

We shall use this implicitly in later analysis. By way of analogy, ordinary light can be considered to be a superposition of planar waves, and speech signals are the result of a speaker with some added background noise. This is illustrated in Figure 2.8 where the (magnitude of the) FTs of a pulse and a ramp are plotted independently in Figures 2.8(d) and (e), and jointly in Figure 2.8(f). The difference between them shown in Figure 2.8(g) is uniformly zero as in Equation 2.15. We can use this to separate a recording of a voice from a recording of its background because the two are added together; later we shall use this to separate two images so we can see the things we are interested in. Some of the later operations apply only for linear systems, and because the FT is linear we can use it there. If the signals were combined by multiplication, it would be a nonlinear process and we could not use the later processes.

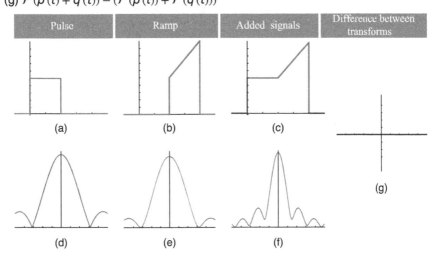

FIGURE 2.8 Superposition and the Fourier transform: (a) $p\,(t)$; (b) $q\,(t)$; (c) $p\,(t) + q\,(t)$; (d) $|\mathcal{F}\,(p\,(t))|$; (e) $|\mathcal{F}\,(q\,(t))|$; (f) $|\mathcal{F}\,(p\,(t) + q\,(t))|$; (g) $\mathcal{F}\,(p\,(t) + q\,(t)) - (\mathcal{F}\,(p\,(t)) + \mathcal{F}\,(q\,(t)))$

2.2.2 Time shift

If a signal is delayed by τ seconds to become $p\,(t - \tau)$ then its FT is defined as

$$\mathcal{F}\,(\text{shifted signal}) = \mathcal{F}\,(p\,(t - \tau)) = Fp\,(\omega) \times e^{-j\omega\tau} \qquad (2.16)$$

Considering the magnitude only

$$|\mathcal{F}\,(p\,(t - \tau))| = |Fp\,(\omega)| \times |e^{-j\omega\tau}|$$

and because $|e^{-j\omega\tau}| = 1$

$$|\mathcal{F}\,(p\,(t - \tau))| = |Fp\,(\omega)|$$

This is known as *shift invariance*: the magnitude of its transform does not change when a signal is shifted in time. The same magnitude is shown in Figure 2.8(d) and in Figure 2.2(b) even though the pulses occur at different times. When processing the transform of recorded speech, it does not matter when a person starts speaking, so long as one is only interested in the magnitude of the transform. This is illustrated in Figure 2.9 where Figure 2.9(a) shows a ramp signal; Figure 2.9(b) a ramp shifted by 2 s; Figure 2.9(c) shows the magnitude of the transform of either ramp; and Figure 2.9(d) the difference between the magnitudes of the transforms of the two ramp functions. The differences in Figure 2.9(d) are very small, O(10^{-16}) where O () represents 'order of' called 'Big Oh', and are because of arithmetic error (in Matlab `sin(pi)` = 1.2246e-16, not zero as we don't have an infinite number of bits) – shift in time does not affect the magnitude of the FT.

FIGURE 2.9 Illustrating shift invariance: (a) ramp; (b) shifted ramp; (c) $|\mathcal{F}(\text{ramp})|$ or $|\mathcal{F}(\text{shifted ramp})|$; (d) $|\mathcal{F}(\text{ramp})| - |\mathcal{F}(\text{shifted ramp})|$

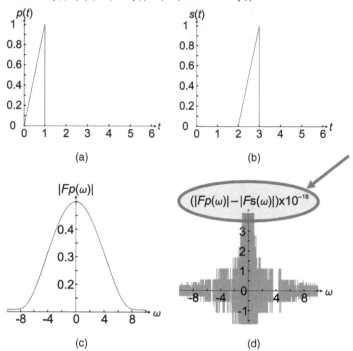

This was also implicit in Figure 2.3 but not mentioned there because we had other things on our mind. In Figure 2.3 the magnitude of the transform of the shifted pulse is unchanged whereas the phase changes from being purely real in Equation 2.5 to the ramp function in Figure 2.3(c). The shift was introduced so that the phase could be visible at that point.

2.2.3 Scaling in time

The *scaling* property concerns what happens in the frequency domain when time is stretched or compressed. Given the inverse relationship between frequency and time, intuitively when time is stretched then frequency is compressed and vice versa. Analytically,

$$\mathcal{F}(\text{stretched signal}) = \mathcal{F}(p(\lambda t)) = \frac{1}{\lambda}Fp\left(\frac{\omega}{\lambda}\right) \qquad (2.19)$$

so a linear scaling of time leads to a linear scaling in frequency. This has previously been seen in Figure 2.2(b) where a lengthy pulse led to fewer frequency components whereas a short pulse has a greater range of frequency components. The derivation of this is,

by Fourier, Equation 2.1　　　$\mathcal{F}\left(p\left(\lambda t\right)\right) = \displaystyle\int_{-\infty}^{\infty} p\left(\lambda t\right) e^{-j\omega\lambda t} dt$

By scaling: $\tau = \lambda t$ and $\omega' = 2\pi/\tau$　　　$= \displaystyle\int_{-\infty}^{\infty} p\left(\tau\right) e^{-j\omega'\tau} dt$

and because $d\tau/dt = \lambda$ so $dt = d\tau/\lambda$　　$= \dfrac{1}{\lambda} \displaystyle\int_{-\infty}^{\infty} p\left(\tau\right) e^{-j\omega'\tau} d\tau$

thus　　　　　　　　　　　　　　$= \dfrac{1}{\lambda} Fp\left(\omega'\right)$

and because $\omega' = 2\pi/\tau = \omega/\lambda$　　$= \dfrac{1}{\lambda} Fp\left(\dfrac{\omega}{\lambda}\right)$

2.2.4　Parseval's theorem (Rayleigh's theorem)

Parseval's theorem largely states that the total energy within a signal is the same in both the time and in the frequency domains (which one would anticipate from the conservation of energy and because the FT changes the representation only, not the content). It actually exists for Fourier series. *Rayleigh's theorem* (or even Rayleigh's identity) is its counterpart for the FT. Graphically, the area under the two curves is the same, which can be inferred from the FT of a Gaussian signal. This is also why the FT is a *unitary* operator. The energy of $p\left(t\right)$ is given by $|p\left(t\right)|^2$ so Rayleigh's theorem states that

$$\int_{-\infty}^{\infty} |p\left(t\right)|^2 dt = \frac{1}{2\pi} \int_{-\infty}^{\infty} |Fp\left(\omega\right)|^2 d\omega \qquad (2.20)$$

In maths, Parseval's theorem is known as *Plancherel's theorem*. So we have a trichotomy: which name should we use? The term in most common use in Engineering is Parseval's theorem, so we shall call it that from now on. To derive it, we need to consider the real and imaginary parts of a signal and from Equation 2.7

$$|p\left(t\right)|^2 = \left(\text{Re}\left(p\left(t\right)\right)\right)^2 + \left(\text{Im}\left(p\left(t\right)\right)\right)^2$$

by completing the square $= \left(\text{Re}\left(p\left(t\right)\right) + j\text{Im}\left(p\left(t\right)\right)\right)\left(\text{Re}\left(p\left(t\right)\right) - j\text{Im}\left(p\left(t\right)\right)\right)$

because $\overline{p\left(t\right)}$ is the *complex conjugate* of $p\left(t\right)$ we can simplify this as

$$|p\left(t\right)|^2 = p\left(t\right) \times \overline{p\left(t\right)}$$

which gives
$$\int_{-\infty}^{\infty} |p(t)|^2 dt = \int_{-\infty}^{\infty} p(t) \times \overline{p(t)} dt$$

by the inverse FT
$$= \int_{-\infty}^{\infty} p(t) \left(\overline{\frac{1}{2\pi} \int_{-\infty}^{\infty} Fp(\omega) e^{j\omega t} d\omega} \right) dt$$

so, by conjugation
$$= \int_{-\infty}^{\infty} p(t) \frac{1}{2\pi} \int_{-\infty}^{\infty} \overline{Fp(\omega)} e^{-j\omega t} d\omega dt$$

by re-ordering
$$= \int_{-\infty}^{\infty} \overline{Fp(\omega)} \frac{1}{2\pi} \int_{-\infty}^{\infty} p(t) e^{-j\omega t} dt d\omega$$

so
$$= \frac{1}{2\pi} \int_{-\infty}^{\infty} \overline{Fp(\omega)} \times Fp(\omega) \, d\omega$$

which gives
$$= \frac{1}{2\pi} \int_{-\infty}^{\infty} |Fp(\omega)|^2 d\omega$$

2.2.5 Symmetry

The FT is intrinsically symmetric because of its basis in positive and negative frequencies. When manipulating the FT it can be convenient to split the transform into its *even* and *odd* parts, which are the real and imaginary parts

$$\text{even}(Fp(\omega)) = Fp_e(\omega) = \text{Re}(Fp(\omega)) \tag{2.21}$$

$$\text{odd}(Fp(\omega)) = Fp_o(\omega) = \text{Im}(Fp(\omega)) \tag{2.22}$$

and if the real part of the FT is based on cosine functions and the imaginary part is based on sine functions, then

$$Fp_e(\omega) = Fp_e(-\omega) \tag{2.23}$$

$$Fp_o(\omega) = -Fp_o(-\omega) \tag{2.24}$$

By Equation 2.23 the even part is symmetric, whereas by Equation 2.24 the odd part is antisymmetric. This shows the symmetry in magnitude and phase which is

$$|Fp(\omega)| = |Fp(-\omega)| \tag{2.25}$$

$$\arg(Fp(\omega)) = -\arg(Fp(-\omega)) \tag{2.26}$$

which can be observed in Figure 2.3.

The FT is then

$$Fp(\omega) = Fp_e(\omega) + Fp_o(\omega) \tag{2.27}$$

By Equations 2.23 and 2.24 $= Fp_e(-\omega) - Fp_o(-\omega)$

By conjugation $= \overline{Fp_e(-\omega) + Fp_o(-\omega)}$

So the *symmetry* of the continuous FT is then

$$Fp(\omega) = \overline{Fp(-\omega)} \tag{2.28}$$

which arises from the symmetry of the basis functions

$$e^{j\omega t} = \overline{e^{-j\omega t}}$$

2.2.6   Differentiation

The (Fourier) *differentiation theorem* states that the FT of the differentiated function

$$\mathcal{F}\left(\frac{dp(t)}{dt}\right) = \mathcal{F}\left(p'(t)\right) = j\omega Fp(\omega) \tag{2.29}$$

This is then considerably easier in the frequency domain because it just requires multiplication of the transform by $j\omega$. To demonstrate this, by the inverse FT (Equation 2.9) we have

$$\frac{d}{dt}(p(t)) = \frac{d}{dt}\left(\frac{1}{2\pi}\int_{-\infty}^{\infty} Fp(\omega)e^{j\omega t}d\omega\right)$$

via-Leibniz' rule $= \frac{1}{2\pi}\int_{-\infty}^{\infty} j\omega Fp(\omega)e^{j\omega t}d\omega$

$$= \mathcal{F}^{-1}\left(j\omega Fp(\omega)\right)$$

which is another form of Equation 2.29.

2.2.7   Uncertainty principle

In Equation 2.14 we showed that there is an inverse relationship between the variance of a Gaussian function g and its transform $\mathcal{F}(g)$, namely that variance $(\mathcal{F}(g)) = 1/\text{variance}(g)$. Now if the variance of the time-domain function tends to zero, then the variance of the frequency-domain version tends to infinity: one cannot simultaneously minimise the variance in the two domains. This implies that an infinitely short pulse is represented by an infinite set of frequencies (as shown for a delta function previously). This means that when we are storing music, short pulses require a vast range of frequencies: there is a tradeoff between resolution in the time domain and resolution in the frequency domain. This is the *uncertainty principle,* which can be expressed using the variance of the two signals' variance () as

$$\text{variance}\left(\mathcal{F}\left(f(t)\right)\right) \times \text{variance}\left(f(t)\right) = 1 \qquad (2.30)$$

which emphasises the relationship between the time and the frequency domains. This can be extended to Heisenberg's uncertainty principle in quantum mechanics where a particle's position and momentum cannot be known exactly, at the same time. Given that particles are waves, this translates to the Fourier uncertainty principle.

2.2.8 Modulation

Modulation is a process used in the transmission of radio signals, so that we can listen to lots of different radio stations. If, say, speech occupies a limited range of frequencies (or bandwidth) and a lot of radio is speech or music, then the speech/music can be modulated using a carrier signal before transmission. After the signal is received it is demodulated and the speech/music is recovered (and listened to). In *amplitude modulation* (AM), the carrier is a signal of frequency ω_0 which modulates the signal which is to be carried $p(t)$. This is achieved by multiplying the two signals. This is shown in Figure 2.10 where Figure 2.10(a) is an original signal, Figure 2.10(b) is the carrier $\cos(\omega_0 t)$ where $\omega_0 = 20$ rad/s and Figure 2.10(c) is the modulated signal, carried by the higher frequency signal.

By Figure 2.10(c), the waveform does not help the understanding here. It's a squiggle! It is much easier to see this in the frequency domain, because the *modulation theorem* states that the FT of a modulated signal $\mathcal{F}\left(p(t)\cos\left(\omega_0 t\right)\right)$ is given by

$$\mathcal{F}\left(p(t)\cos\left(\omega_0 t\right)\right) = \frac{1}{2}Fp\left(\omega - \omega_0\right) + \frac{1}{2}Fp\left(\omega + \omega_0\right) \qquad (2.31)$$

which means that the spectrum of the modulated signal is repeated at the carrier frequency. This is shown in Figure 2.11 and the modulated signal is in Figure 2.11(c). Here, the spectrum of Figure 2.11(a) is repeated at the frequencies of Figure 2.11(b) (which shows the modulation frequency $\omega_0 = 20$). In this way, the FT shows us more clearly how modulation operates than can be seen in the time domain. Now the radio spectrum can be used more efficiently because different stations can use different carrier frequencies (which is why you have to tune the system to pick up your favorite station). OK, it's all digital now and there is *frequency modulation* (FM) – for better performance in respect of fidelity and reduction of noise (and for tolerating multipath effects, though it will be discontinued soon in the UK in favour

FIGURE 2.10 Modulation in the time domain: (a) original signal; (b) carrier; (c) modulated signal

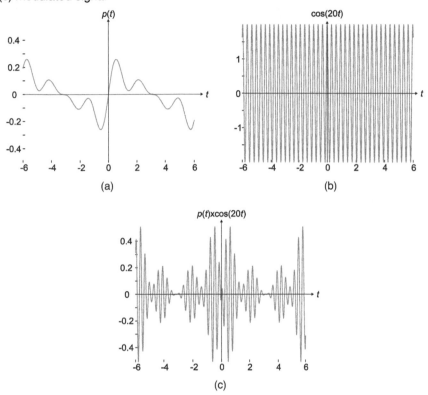

(a)

(b)

(c)

of digital audio), but we won't go there as we're at the start line and not at the chequered flag.

Let us derive it. Part of the process is the *frequency shift theorem*

$$\mathcal{F}^{-1}\left(\text{shifted frequency}\right) = \mathcal{F}^{-1}\left(Fp\left(\omega - \omega_0\right)\right) = p\left(t\right) \times e^{-j\omega_0 t} \quad (2.32)$$

which is the frequency-domain counterpart of the time shift theorem. This suggests that a shift in frequency (modulation) does not change the magnitude of the time-domain signal. This is useful to know as music would be hard to listen to without it. Now for modulation. by definition

$$\mathcal{F}\left(p\left(t\right)\cos\left(\omega_0 t\right)\right) = \int_{-\infty}^{\infty}\left(p\left(t\right)\cos\left(\omega_0 t\right)\right)e^{-j\omega t}dt$$

FIGURE 2.11 Modulation in the frequency domain: (a) original signal spectrum; (b) carrier spectrum; (c) modulated signal spectrum

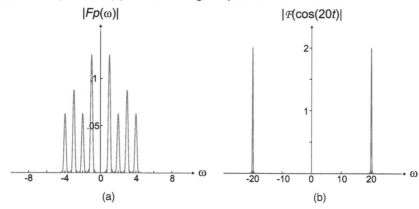

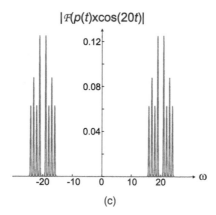

| substituting for $\cos(\omega_0 t)$ | $= \int_{-\infty}^{\infty} \left(p(t) \left(\dfrac{e^{j\omega_0 t} + e^{-j\omega_0 t}}{2} \right) \right) e^{-j\omega t} dt$ |

by rearrangement

$$= \frac{1}{2} \int_{-\infty}^{\infty} p(t)\, e^{j\omega_0 t} e^{-j\omega t} dt + \frac{1}{2} \int_{-\infty}^{\infty} p(t)\, e^{-j\omega_0 t} e^{-j\omega t} dt$$

and simplification

$$= \frac{1}{2} \int_{-\infty}^{\infty} p(t)\, e^{-jt(\omega-\omega_0)t} dt + \frac{1}{2} \int_{-\infty}^{\infty} p(t)\, e^{-jt(\omega+\omega_0)t} dt$$

Giving

$$= \frac{1}{2} Fp\,(\omega - \omega_0) + \frac{1}{2} Fp\,(\omega + \omega_0)$$

This is the spectrum shown in Figure 2.11(c). Note that the modulation theorem is the dual of the time shift theorem because the multiplication is in the time domain, giving shift in the frequency domain.

2.3   PROCESSING SIGNALS USING THE FT

2.3.1   Convolution

Convolution is a process that can be used to determine the relation-ships between signals in linear time-invariant systems. It is often used to calculate the output of a system given an input and a system response function. It is defined as

$$p(t) * q(t) = \int_{-\infty}^{\infty} p(\tau) q(t - \tau) \, d\tau \qquad (2.33)$$

where $*$ denotes convolution. Essentially, $q(t - \tau)$ describes the *memory function* of a system, its basic response to an input. When an input signal and a memory function are multiplied and added over all time we calculate the response of a system via the convolution integral.

The convolution process is illustrated in Figure 2.12. Here we have a pulse in Figure 2.12(a) and an exponential response in Figure 2.12(b). The exponential response is equivalent to a system output that decays with time. We shall find what system the out-put derives from after this, for now it is just an exponentially decreasing output. Convolution is phrased in terms of a pseudo-variable τ because we need to be able to reverse the time-domain functions; time itself cannot be shifted (as we need in the convolution process), whereas τ can be. When the exponential function is reversed in time τ, to form $q(t - \tau)$ then it becomes a mem-ory function because its value describes how much is remembered at that point, Figure 2.12(c). ('Convolution' derives in part from the Latin *volvere*, which means to twist or turn about; *convolvo* = I convolve!) Convolution is a process integrating the result of multiplication whilst the memory function is shifted in time. Three points at different times (including the maximum) and the convolution result are shown in Figure 2.12(d). At each point, the result of the integration is the area under the curve (from the origin). So the area of intersection starts by increasing, then reaches a maximum and thereafter reduces exponentially.

$*$ is not used the same way as in multiplication (in software) or as it is when texting or to censor swearwords!

Mathematically, the pulse in Figure 2.12(a) is

$$p(t) = \begin{cases} 1 & 0 \leq t \leq T \\ 0 & \text{otherwise} \end{cases} \qquad (2.34)$$

The exponential function in Figure 2.12(b) is

$$q(t) = \begin{cases} e^{-at} & t \geq 0 \\ 0 & t < 0 \end{cases}$$

FIGURE 2.12 Convolution: (a) pulse $p(t)$; (b) exponential response $q(t)$;
(c) memory function $q(t-\tau)$; (d) result of convolution $p(t) * q(t)$ and its derivation

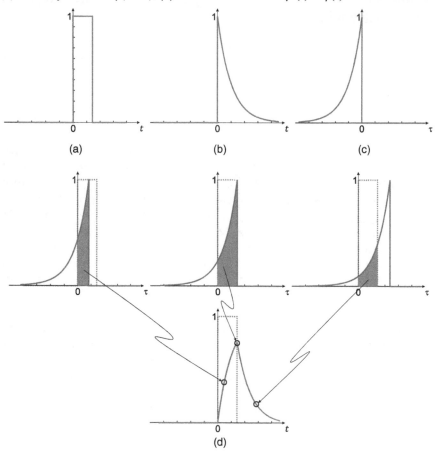

which is the solution to the first-order differential equation (a simpler derivation of this is by signal processing theory in Section 5.5.1; we shall return to this differential equation in Section 2.6.1).

$$\frac{dq}{dt} + aq = 0 \qquad (2.35)$$

The memory function in Figure 2.12(c) is then

$$q(t-\tau) = \begin{cases} 0 & \tau > 0 \\ e^{a\tau} & \tau \leq 0 \end{cases}$$

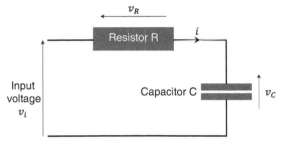

FIGURE 2.13 Circuit producing the output in Figure 2.12(d)

The convolution output in Figure 2.12(d) is

$$p(t) * q(t) = \begin{cases} \displaystyle\int_0^t p(\tau)\, q(t-\tau)\, d\tau & \tau \geq 0 \\ 0 & \tau < 0 \end{cases} \tag{2.36}$$

which results from the product of an input and how much of it can be remembered and this gives the waveform shown. The convolution of the pulse $p(t)$ with the exponential decay $q(t)$ rises and falls in an exponential manner. This is maths, now let us look at a real system.

Now for a bit of electronics! The exponential function is actually the output of an RC network shown in Figure 2.13, the voltage across the capacitor v_C. In this circuit, the capacitor either charges or discharges through the resistor. When a voltage is applied at the input, the capacitor aims to charge to that voltage (it's like filling up a bucket with water and the flow of water is the current i). When no voltage is applied at the input, the capacitor discharges through the resistor (only if it was previously charged, it can't discharge from nothing). The rate of charging – or discharging (emptying said bucket) – is controlled by the time constant, which is a product of the resistance (R) and the capacitance (C). If R and C are large it will take a longer time to charge/discharge whereas if both are small then the charging/discharging will be fast.

It's a bit like a student drinking beer at the Students' Union bar: the capacitance is the amount of beer in the drinker's belly and the resistance is the attractiveness of whosoever they are trying to chat up. The time constants for 'charging' and discharging beer are not the same!

Now Figure 2.12(d) is actually the output of an RC network with a pulse applied at the input. The shape of the underlying pulse is shown as dotted to show the construction of the final output waveform. The output starts with the capacitor charging when the pulse is applied, and the capacitor discharges when the pulse ceases. The capacitor charges when the pulse exists, so the voltage across it rises; when the pulse stops the capacitor stops charging and discharges through the resistor.

In both stages, the exponential function is described by the time constant RC. So the convolution function, Figure 2.12(d), calculates the
response of the RC network when a pulse is applied at its input. We
could do this using maths specific to the circuit as an extension to the
previous analysis. Here we need an input and a system response function and then calculate the output using convolution. Convolution is
a general and very useful function indeed. We shall meet it again and
again in this book.

In electronics, the network is actually a low-pass filter because at
high frequencies the impedance of the capacitor is zero, so it is only the
low frequencies that pass through the circuit. It could be part of the
crossover network in a sound system to select the music that is fed to
the bass speakers (which are the big ones that boom not tweeters that,
well, tweet). We shall hear more about filters (and sound systems) later
on in Section 2.6 and cover the input and system response functions in
more detail.

There is an alternative way to calculate convolution. Using the FT,
the transform of the convolution operation is

$$\mathcal{F}\left(p\left(t\right) * q\left(t\right)\right) = \mathcal{F}\left(\int_{-\infty}^{\infty} p\left(\tau\right) q\left(t - \tau\right) d\tau\right)$$

by the FT, $\qquad \mathcal{F}\left(p\left(t\right) * q\left(t\right)\right) = \int_{-\infty}^{\infty}\left\{\int_{-\infty}^{\infty} p\left(\tau\right) q\left(t - \tau\right) d\tau\right\} e^{-j\omega t} dt$
Equation 2.1

collecting similar $\qquad = \int_{-\infty}^{\infty}\int_{-\infty}^{\infty} q\left(t - \tau\right) e^{-j\omega t} dt\, p\left(\tau\right) d\tau$
terms

as the inner term is $\qquad = \int_{-\infty}^{\infty} \mathcal{F}\left(q\left(t - \tau\right)\right) p\left(\tau\right) d\tau$
an FT

by substitution from $\qquad = \mathcal{F}\left(q\left(t\right)\right) \times \int_{-\infty}^{\infty} p\left(\tau\right) e^{-j\omega\tau} d\tau$
Equation 2.16

so $\qquad = \mathcal{F}\left(q\left(t\right)\right) \times \mathcal{F}\left(p\left(t\right)\right)$

So in the frequency-domain, convolution is the result of multiplying
the two transforms and this is called the *convolution theorem*

$$\mathcal{F}\left(p\left(t\right) * q\left(t\right)\right) = \mathcal{F}\left(p\left(t\right)\right) \times \mathcal{F}\left(q\left(t\right)\right) \qquad (2.37)$$

and this is stated in key point 3.

The dual of convolution in the time domain is multiplication in the frequency domain, and vice versa.

This applies in reverse too: convolution in frequency is equivalent to multiplication in time. This is important for Fourier itself: the FT is the result of convolving sine waves of differing frequency with a signal.

Via the convolution theorem, the convolution of two signals can be achieved using the inverse FT as

$$p(t) * q(t) = \mathcal{F}^{-1}(\mathcal{F}(p(t)) \times \mathcal{F}(q(t))) \qquad (2.38)$$

This gives an alternative way to calculate the convolution integral. We shall later find this employed with dramatic effect to speed up algorithms, especially in 2-D image analysis. Note that convolution is *commutative* and so order does not matter:

$$p(t) * q(t) = q(t) * p(t) \qquad (2.39)$$

Largely by this, convolution is also *associative* in that

$$p(t) * (q(t) * r(t)) = (p(t) * q(t)) * r(t) \qquad (2.40)$$

and finally it is *distributive*

$$p(t) * (q(t) + r(t)) = p(t) * q(t) + p(t) * r(t) \qquad (2.41)$$

These imply that there is a natural algebra in convolution, which one expects in linear systems.

2.3.2 Correlation

Correlation is a process that determines relationships between signals and is defined as

$$p(t) \otimes q(t) = \int_{-\infty}^{\infty} p(\tau) q(t + \tau) d\tau \qquad (2.42)$$

where $\otimes$ denotes correlation. Essentially, when a signal is multiplied by one that has been shifted over all time and the multiplication result is added then we calculate the relationship between the two signals. It is a form of matching and correlation is a measure of the similarity between two signals. There are alternative symbols that can be used for correlation such as $\odot$ but none appears to have found common use (the latter symbol looks like a target anyway – it is also the symbol for the Sun in Egyptian hieroglyphics and the end of a trail to Boy Scouts – so we shall not use it here).

We shall not use any circuits to explain correlation, because it has a simple analogy: it is equivalent to matching two signals. If the time axis

FIGURE 2.14 Correlation: (a) $p(t)$; (b) $q(t)$; (c) result of correlation $p(t) \otimes q(t)$ and its derivation

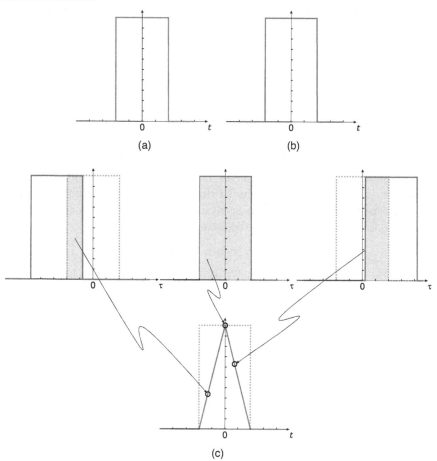

(a)   (b)

(c)

is not reversed as it is in convolution, then correlation is the process of passing one signal across another, and measuring the area of intersection between the two. This is shown in Figure 2.14 where we match a pulse to itself. In the process of passing one pulse across the other, initially the area of intersection is small and rises linearly until a maximum when the two pulses are superimposed, before decreasing. Three points showing three different amounts of correlation are shown (including the maximum). The shape of the underlying pulse is dotted to show the construction of the final output waveform. Mathematically, the process is the integration of a constant function so the correlation function has slope $\pm t$. If the two waveforms are mismatched, then the correlation is low; if the two are similar, then the correlation function peaks, as it does here. The process is actually equivalent to the one used in the reception of the stream of binary signals in a Web connection over, say, a noisy phone line. There, the sequence of binary symbols is contaminated by noise, which corrupts any communication channel. If one were to threshold the signal, detecting a '1' when it is above the chosen threshold and a '0' otherwise, detection would be prone to noise. Instead, if the two are correlated in a matched filter, then detection of a bit depends on whether a '0' or a '1' is more likely. This is implemented by an 'integrate and dump' receiver, and the integration is precisely the $\pm t$ in the correlation function in Figure 2.14. This will be met again in Section 4.4.3.

Figure 2.14 actually illustrates a process called *autocorrelation*, because we are correlating a pulse with itself. When bits are received in a communications channel the process is correlation because a template bit is matched to the incoming signal. The *Wiener-Khinchin theorem* describes autocorrelation because it is a special case of correlation where $p(t) = q(t)$ as

> Wiener – the father of Cybernetics no less – must have been one of the most forgetful engineers ever. In one apocryphal tale, he was on his way home but in some difficulty so he asked a young chap 'Excuse me young man, do you know where my home is?'. 'Yes, Dad...' came the reply.

$$p(t) \otimes p(t) = |\mathcal{F}(p(t))|^2 \qquad (2.43)$$

and the autocorrelation function describes the *energy* in the signal.

As with convolution, there is a frequency-domain implementation of correlation. Because the difference between Equation 2.33 and Equation 2.42 is the time inversion where $q(t - \tau)$ becomes $q(t + \tau)$, convolution can be achieved via Equation 2.37 by inverting $q(t - \tau)$ along the time axis. In the frequency-domain correlation is thus

$$\mathcal{F}(p(t) \otimes q(t)) = \mathcal{F}(q(-t)) \times \mathcal{F}(p(t)) \qquad (2.44)$$

and the implementation is

$$p(t) \otimes q(t) = \mathcal{F}^{-1}(\mathcal{F}(q(-t)) \times \mathcal{F}(p(t))) \qquad (2.45)$$

This can also be achieved, for real-valued signals, using conjugation as

$$p(t) \otimes q(t) = \mathcal{F}^{-1}\left(\overline{\mathcal{F}}(q(t)) \times \mathcal{F}(p(t))\right) \qquad (2.46)$$

where $\overline{\mathcal{F}}(q(t))$ denotes the complex conjugate of $\mathcal{F}(q(t))$. There is no 'correlation theorem': the 'convolution theorem' describes both processes. Note that if even functions are symmetric around $t = 0$, then convolution and correlation are the same for even signals. Also, unlike convolution, correlation is neither commutative nor associative. By linearity, it is distributive, as per Equation 2.41 (for correlation).

2.4 WHAT IS THE IMPORTANCE OF PHASE?

We last encountered phase in Section 2.1.2, and have not considered it again (especially in Section 2.2.2). Let us rectify that now. (This is consistent with many books on this subject – a lot of understanding can be achieved by magnitude alone. We are not criticising here ... well, actually we are, read on! There again, there has been previous emphasis on the importance of phase Oppenheim & Lim, 1981.) By Equation 2.6 phase is an integral part of the FT and phase is implicit in the transform; a signal cannot be reconstructed by the inverse FT without its phase. So far, signals have been viewed in terms of their magnitude, largely since that is what was needed at the time: if a signal is made up as a collection of sine waves, then to understand the process, it is sufficient to know the magnitude (how much) of the signals that make up a complex signal. The phase is important when it is necessary to know when the signals occurred, rather than just the magnitude information. We shall show the phase by one property that has already been shown, where phase was noted as of importance, and one application where phase is implicit and has important properties.

By further analysis of Equation 2.6, we can represent the real and the imaginary parts as

$$Fp\,(\omega) = \text{Re}\,(Fp\,(\omega)) + j \times \text{Im}\,(Fp\,(\omega))$$
$$= |Fp\,(\omega)| \times \cos\,(\arg\,(Fp\,(\omega))) + j\,|Fp\,(\omega)| \times \sin\,(\arg\,(Fp\,(\omega))) \tag{2.47}$$

Clearly, the phase is part of this equation, but one question is how important is it? Let us find out.

2.4.1 Phase in signal reconstruction

In order to reconstruct a signal, we take the inverse FT of a signal as

$$p\,(t) = \mathcal{F}^{-1}\,(Fp\,(\omega))$$
$$= \mathcal{F}^{-1}\,(|Fp\,(\omega)| \times \cos\,(\arg\,(Fp\,(\omega))) + j\,|Fp\,(\omega)| \times \sin\,(\arg\,(Fp\,(\omega))))$$

For two signals p and q, with FTs Fp and Fq we shall swap the phase between them and see what we get. We shall label the results of swapping the phases as A and B where

$$A = \mathcal{F}^{-1}\,(|Fp\,(\omega)| \times \cos\,(\arg\,(Fq\,(\omega))) + j\,|Fp\,(\omega)| \times \sin\,(\arg\,(Fq\,(\omega))))$$
$$B = \mathcal{F}^{-1}\,(|Fq\,(\omega)| \times \cos\,(\arg\,(Fp\,(\omega))) + j\,|Fq\,(\omega)| \times \sin\,(\arg\,(Fp\,(\omega))))$$

The effects of swapping the phase are shown in Figure 2.15. Here, we have two different signals p and q shown in Figures 2.15(a) and (b). The reconstruction resulting from the magnitude of p and the phase of q is shown in Figure 2.15(c) and the reconstruction

FIGURE 2.15 Swapping the phase in reconstruction: (a) $|p(t)|$; (b) $|q(t)|$; (c) $A = \mathcal{F}^{-1}(|Fp| \times \text{arg}(Fq))$; (d) $B = \mathcal{F}^{-1}(|Fq| \times \text{arg}(Fp))$

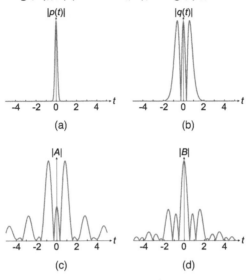

resulting from the magnitude of q and the phase of p is shown in Figure 2.15(d). The actual size of the signals is not shown and this naturally follows the magnitude spectrum; what is of interest here is the overall shape. The phase dominates the reconstruction process because the reconstructed signal is more similar to the signal whose phase was used in reconstruction, rather than to the one whose magnitude was used. The sidelobes in Figure 2.15(c) are similar to those of Figure 2.15(b) and the relative absence is shown in Figures 2.15(a) and (d). The signals are not a perfect match and that should be expected. This will become (visually) clearer when we consider transforms of images, in Section 4.3.5 (if you're unconvinced – which is not unreasonable, as it's a bit of a stretch here – go and have a look at that!). Later we shall find a computer vision operator which is predicated by the existence of phase, Section 6.5.

2.4.2 Phase in shift invariance

Previously, in Section 2.2.2 when we considered shift invariance, we showed that the magnitude of the FT does not change when a signal is shifted in time. We cannot get something for nothing as that would violate the first law of thermodynamics. What has changed is the phase. We repeat the ramp and its shifted version in Figures 2.16(a) and (c), respectively. The phase of each signal is shown in Figures 2.16(b) and (d). Clearly the phase is different. Phase can be important, strike two! We appreciate that the argument of whether

FIGURE 2.16 Phase in shift invariance: (a) ramp (Figure 2.9a); (b) phase of
$\mathcal{F}$ (ramp); (c) shifted ramp; (d) phase of $\mathcal{F}$ (shifted ramp)

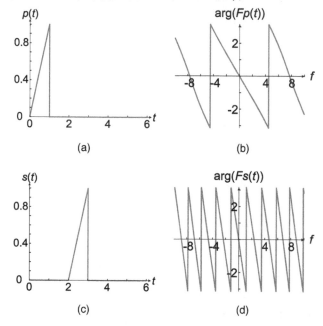

phase is more important than magnitude is a rather arbitrary one. One
does not exist without the other. Here we are simply demonstrating
that phase should not be ignored.

2.5 WINDOWING THE FT DATA

2.5.1 Basic windowing

If we seek to analyse a particular section of a waveform, then the process is known as *windowing*. The most straightforward way to define
the function is simply to set the window function as unity during the
time of interest, and zero elsewhere. The product of the window and
the waveform then results in the section of interest. We can define a
window function wR as

$$wR\,(t) = \begin{cases} 1 & -3T/2 \leq t < 3T/2 \\ 0 & \text{otherwise} \end{cases} \qquad (2.50)$$

where T denotes a period as usual, and wR is described as a *rectangular
window* because of its shape. This is shown applied in Figure 2.17, and
clearly samples the function in the time domain. Windowing is used
within the *Short-Time Fourier transform* (STFT) which is described for

FIGURE 2.17 Windowing functions (in the time domain) and their effects on frequency: (a) sine wave; (b) delta functions; (c) rectangular window; (d) magnitude of sinc function; (e) windowed sine wave; (f) transform of windowed sine wave (= transform of sine wave convolved with sinc function); (g) collection of sine waves; (h) Fourier transform of windowed collection of sine waves

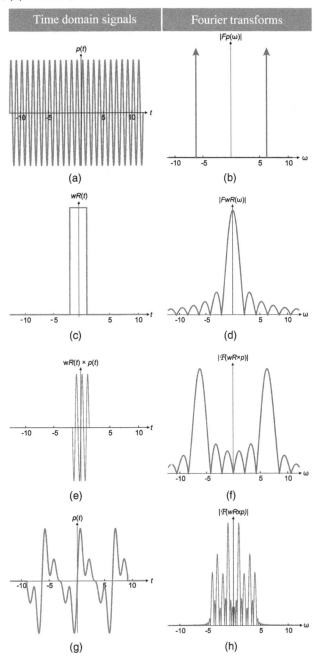

a window function w as

$$STFp\,(\omega;\tau) = \int_{-\infty}^{\infty} p\,(t)\,w\,(t-\tau)\,e^{-j\omega t}dt \qquad (2.51)$$

which can be used to decompose the transform into computation from periods of differing time. However, a rectangular window also has some rather nasty properties (it adds some sediment to our beer). To understand these additional and undesirable properties we must remember from Section 2.3.1 that multiplication in the time domain is convolution in the frequency domain. Applying the windowing in the time domain is by multiplication so we will need to convolve spectra. Now we have seen the FT of the pulse before in Section 2.1.1, it is the sinc function, Figure 2.17(d). The sinc function has *sidelobes* – those bits where the transform bounces – and these extend to infinity. So if we multiply a function in the time domain, Figure 2.17(a), by a rectangular window Figure 2.17(b), this is equivalent in the frequency domain to convolving the transform of the signal with the transform of the rectangular window. This means that the FT now repeats to infinity in the frequency domain, Figure 2.17(f), whereas there was only one frequency in the windowed time-domain signal, Figure 2.17(e). It is actually because of the implausibly sharp sides of the windowing function (where the rectangular function changes from 0 to 1 – or vice versa – in zero time). If something changes very quickly, this implies high frequency. These are the high frequencies we have added by the windowing process.

Figure 2.17(a) shows a sine wave of frequency 1 Hz, period $T = 1$ s; $\omega_0 = 2\pi$ rad/s.

$$p\,(t) = \sin\,(2\pi t)$$

From Equation 2.12 the continuous FT of the sine wave, Figure 2.17(b), is a pair of Dirac delta functions, each at an angular frequency 2π radians/s.

$$|Fp\,(\omega)| = \pi\,(\delta\,(\omega - 2\pi) + \delta\,(\omega + 2\pi))$$

A rectangular window $3T$ (s) in duration, $wR\,(t)$, is shown in Figure 2.17(c) and the absolute value of its transform magnitude in Figure 2.17(d). When the rectangular function is used to window the signal by forming the product $wR\,(t) \times p\,(t)$ we find a shortened sine wave Figure 2.17(e). The transform of this shortened signal is shown in Figure 2.17(e) and arises from convolving the spectrum in Figure 2.17(b) with the spectrum in Figure 2.17(d). If there are sidelobes in Figure 2.17(d) then the high frequencies continue in Figure 2.17(f), and these are the frequencies introduced by the windowing function. This makes it less easy to understand the FT when the sine waves of Figure 2.17(b) actually become those of Figure 2.17(f) when a shorter time sample is observed: the simpler Dirac functions are confused by the sinc function with its sidelobes. When the collection of sine waves is windowed with a pulse of longer duration, then the spectrum is of the form shown in Figure 2.17(h). (Note that the spectrum in Figure 2.17(h) does not look as clean as the one originally in Figure 2.1: there are spikes relating to the component frequencies but there is also more noise beneath them.) This illustrates the complexity of the frequency domain. It is certainly very rich, allowing convolution and correlation and other examples that will follow later, but it is also complicated. As we need to work in the frequency domain, we need also to find out what happens with the FT (a side argument is that sampling things – and that is an extended form of windowing – always causes problems, but we need computers too and they work with sampled signals). This windowing section is by way of introduction to sampling, which comes in the next chapter.

We should aim to avoid the effects introduced by the windowing function. To accomplish this, we need to change the shape of the sampling window and its duration. For a periodic function, ideally, the window would be of infinite length and the resulting spectrum would be that of Figure 2.17(a). The transform pair of a uniformly flat function is a spike in Section 2.1.4, so if the window was of infinite duration, the result is equivalent to the convolution of the transform of the sine waves with a single delta function. But a window of infinite duration is not a window at all; a window is, by definition, of limited duration. Naturally, for a signal that is aperiodic, a window is confined to the period of interest. In Figure 2.17(h) we actually used a window of more optimum duration ($wT = 3T$), so that the result looked similar to Figure 2.1. In Fourier what happens is a consequence of what has been defined (naturally) but the result sometimes needs a deeper understanding. First, we shall consider changing the shape of the windowing function.

2.5.2 Hanning and Hamming window operators

The (rather nasty) properties of the rectangular window mean that we need to seek other functions. We have already met the Gaussian function in Section 2.1.5 and that can serve as a window with properties well known in the frequency domain. There are actually many windowing functions, all with different properties (and they are often eponymous). We shall look at some major ones. For a rectangular window centred at the origin, the first place to look is some form of cosine function because it is also centred and symmetric about the origin and more importantly it is smooth (avoiding discontinuities means avoiding spurious high frequencies). The first windowing function comes from the 19th century and is the *Hann window* (from Julius von Hann), which is based around a cosine function and is defined for duration *TW* (centred at the origin) as

$$
wHAN(t; TW) = \begin{cases} 0.5 + 0.5\cos\left(\dfrac{\pi t}{\frac{TW}{2}}\right) & -TW/2 \le t < TW/2 \\ 0 & \text{otherwise} \end{cases} \quad (2.52)
$$

The Hann window is usually described as the *Hanning window*, largely by its linguistic similarity to the Hamming window which comes next. It does of course add confusion, because it is easily misheard – but then they are similar in design and effect, though there are some differences in operation, which we shall describe after the Hamming window, which naturally comes next. By virtue of its formulation, the Hanning window is also known as a *raised cosine window*, because the cosine is raised by 0.5 and because

$$
\frac{1}{2}\left(1 + \cos\left(\frac{\pi t}{\frac{TW}{2}}\right)\right) = \cos^2\left(\frac{\pi t}{TW}\right)
$$

With slight change to the amplitudes of the coefficients of the windowing function, the *Hamming window* (from Richard W. Hamming) arrived later and is

$$
wHAM(t; TW) = \begin{cases} 0.54 + 0.46\cos\left(\dfrac{\pi t}{\frac{TW}{2}}\right) & -TW/2 \le t < TW/2 \\ 0 & \text{otherwise} \end{cases} \quad (2.53)
$$

and this is plotted together with the Hanning window in Figure 2.18(a), both for *TW* = 3*T* (s) in duration, so that the eventual result can be compared with Figure 2.17(h). In the Hamming window the coefficients are chosen so as to cancel the first sidelobe of the Hann window, giving it a height of about one-fifth of the Hann window, and this can be seen when the two transforms are compared in Figure 2.18(b). More precise values of the coefficients give rise to other effects in the frequency domain, in particular that the transform is flatter in that the sidelobes are of the same peak magnitude (known as equi-ripple). The real difference is that the Hamming window reduces more the frequency components that are close to the central peak, whereas the Hanning window reduces more the frequency components that are further from the central peak. By their formulation there is not much difference as in Figure 2.18(b), but this could be critical in sensitive systems. The Hamming window comes from the same chap who originally devised the Hamming distance which is used in error analysis in coding systems. Hamming also worked on the Manhattan project so he had an interesting career

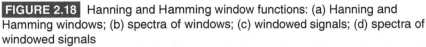

FIGURE 2.18 Hanning and Hamming window functions: (a) Hanning and Hamming windows; (b) spectra of windows; (c) windowed signals; (d) spectra of windowed signals

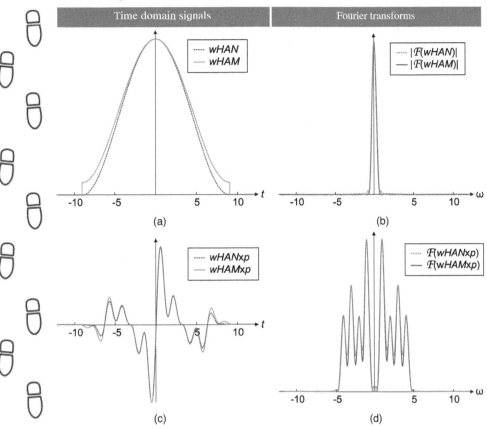

indeed. The Short-Time FT (STFT) is more usually expressed using one of these window functions.

The effects of the two windows on the signal formed by adding up sine waves (from p = 0.5*sin(t)+0.25*sin(2*t)+0.35* sin(3*t)+0.25*sin(4*t)) are shown in Figure 2.18(c) and are shown in the transforms in Figure 2.18(d). These appear much 'cleaner' than the one in Figure 2.17(h) and a reduction in the extra sidelobes can be seen. This is still, however, some difference from Figure 2.1, and that is because of the windowing function. The result in Figure 2.1 arises from calculation that extends much further, though not to infinity. In reality, signals do not extend to infinity (that was part of Fourier's problem when trying to publish his original work, Section 7.1.2) so for periodic signals there will always be a windowing

What's the difference between a Hanning and a Hamming window? It's 0.06 ... baboom ... groan!

function. Since it will be there, we must anticipate its effects, and that is the difference between the result in Figure 2.1 and the result in Figure 2.18(d). Should we increase the duration of the window, then the number of sidelobes will reduce, but they will still be there. These problems will be magnified when we consider sampled signals using a digital computer which we consider in the next chapter.

2.5.3 Window duration

We must now confess to some chicanery. In order to explain window functions, in Figure 2.17 we used an ideal duration to derive the spectral plots (and we did it in Figures 2.1 and 2.11 – guilty m'lord – a classic problem in education is when to skip reality and when to reveal it). This was for understanding, so that the basic message is clear, or that is our aim at least.

When sampling a periodic signal, we used a window which is equivalent to multiplying the signal by a pulse to give the sampled section of the waveform. Noting that the pulse can introduce artefacts, we changed the shape of the windowing function: the Hanning and Hamming windows improved clarity considerably. We have yet to consider the duration of the window function. Naturally, we can use a window of long duration for low frequencies and a short window for high frequencies. The frequency-domain effects introduced by limiting the duration of the window function are fewer with a longer window (because the transform of a longer window tends towards a spike). The effect of changing the duration of a Hanning window is shown in Figure 2.19. We have not shown the (amplitude) scaling function, which here is the area of the windowing function, and for a Hamming window is

$$\text{windowed\_p} = \frac{p(t)\,wHAM(t; TW)}{\left(0.54\left(\frac{TW}{2}\right)\right)}$$

otherwise a longer window would lead to larger estimates of the signal's frequency content. We also show the transform resulting from a window of infinite duration, the FT of the original signal in Figure 2.19(a) which is the set of delta functions in Figure 2.19(b). Here we are defining the length of a window that is appropriate for deriving the FT of a periodic signal. If that signal is of period T s, then Figures 2.19(c), (d) and (e) show the results for window lengths T, $3T$ and $5T$, respectively.

Each of the spectra in Figure 2.19 derives from a time-limited window function of the periodic data and can be considered as derived by convolution of the spectrum of the window function with the spectrum of the signal. The convolution process is implicit in using multiplication by the windowing function: the dual of that multiplication in

FIGURE 2.19 Spectra resulting from Hamming windows of differing duration: (a) time-domain signal, period T; (b) Fourier transform; computed transforms for differing window length TW; (c) $TW = T$; (d) $TW = 3T$; (e) $TW = 5T$

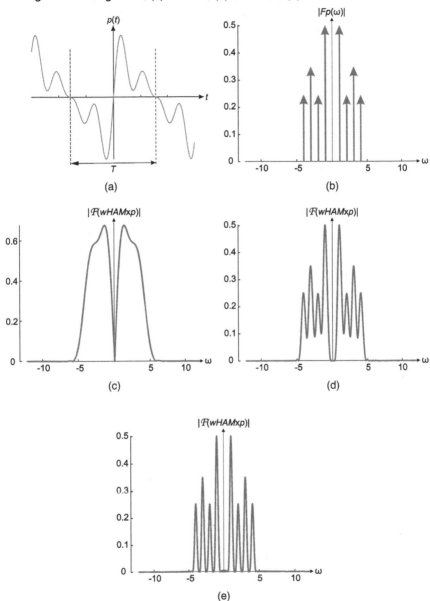

the time domain is convolution in the frequency domain. As the duration of the window function increases, the width of its frequency spectrum decreases and the sidelobes diminish too. So, in Figure 2.19(c) the duration of the window is the shortest so the spectrum of the window function is the widest, and so the spectral components blur into each other. In Figure 2.19(e) the window duration is the longest, so its spectrum is the thinnest and so after the convolution the spectral components are very clear. So, the duration of the window is a compromise and the rule of thumb is in key point 4.

Key point 4

For windowing a periodic signal, choose a window length of three times its period.

If the signal to be windowed is aperiodic, like a section of speech, then a window length is chosen and its effects are noted because that is all we can do. In Figures 2.17(h) and Figures 2.18(d) the square window was of duration $3T$ (used for clarity in the explanation of windowing functions) and the spectral effects of the square window function are improved by the Hanning and Hamming windows (comparing the spectrum in Figure 2.17(h) with that in Figures 2.18(d) and 2.19(d)).

2.5.4 Other windowing functions

In some circumstances it is advantageous to have similar effects on all frequency components, which implies that the transform of the windowing function is as flat as possible. Conversely, it can also be desirable to affect the time-domain components as little as possible, and we shall see that in the flat-top windowing function which comes next. The frequency domain reveals the intimate relationship between magnitude and phase: we cannot have one without the other. This then verges on the domain of filters in signal processing and there are concepts of equal power and linear phase that complicate further the design of the filters (and motivated their many different forms).

A *flat-top window* is a window that is chosen to estimate the amplitude of signals, as opposed to their frequency (this one is not eponymous unless someone had a rather unfortunate surname). It is a formulation of a cosine sum window where the components are $w = \sum_k a_k \cos\left(k\pi t/(TW/2)\right)$. The flat-top window is

$$wFT(t) = \begin{cases} a_0 + a_1 \cos\left(\dfrac{\pi t}{\frac{TW}{2}}\right) + a_2 \cos\left(\dfrac{2\pi t}{\frac{TW}{2}}\right) + \cdots \\ \\ \cdots + a_3 \cos\left(\dfrac{3\pi t}{\frac{TW}{2}}\right) + a_4 \cos\left(\dfrac{4\pi t}{\frac{TW}{2}}\right) \\ \\ 0 \end{cases} \quad \begin{array}{l} -TW/2 \le t \, lt; \, TW/2 \\ \\ \\ \text{otherwise} \end{array}$$

$$(2.54)$$

FIGURE 2.20 Flat-top window function (a) flat-top window *wFT*; (b) spectrum of *wFT*

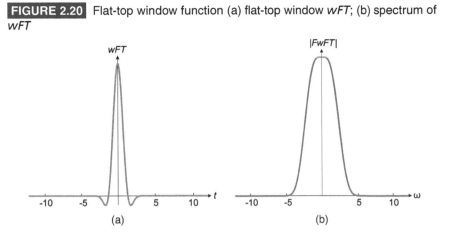

(a)   (b)

where in Matlab's implementation https://uk.mathworks.com/help/signal/ref/flattopwin.html $a_0 = 0.21557895$, $a_1 = 0.41663158$, $a_2 = 0.277263158$, $a_3 = 0.083578947$, and $a_4 = 0.006947368$. There are more flat-top windowing functions because of its popularity [Heinzel et al., 2002] though there appear to be few academic studies. The flat-top window and its transform are shown in Figure 2.20 though the time duration here is chosen for frequency-domain visualisation and, in application, its time duration is usually longer than shown here.

These windows have been used to select parts of the signal that are of interest. This is where the other main form of the FT comes in because when we repeatedly select parts of a signal, we are sampling the signal, as in a computer. This is where we find the most practical form of the FT, because it has many more applications than the continuous version. A more advanced form of windowing is *wavelets* which allow localised frequency analysis at different scales, as covered in Section 5.7.

2.6   FILTERING THE FT DATA

2.6.1   Basic filters and signal processing

2.6.1.1   Low-pass, high-pass and band-pass filters

Filtering is the process by which we select chosen parts of a signal and remove them. This takes our material much closer to *signal processing*, which we shall need for applications later. There are three main types of filter: *low-pass, high-pass* and *band-pass*, which select the ranges of frequencies shown in Figure 2.21. There is also a *band stop* filter,

FIGURE 2.21 Filtering continuous signals

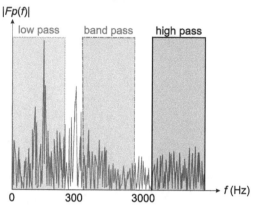

which is (loosely) a combination of low-pass and high-pass filters. This is ideal filtering: selecting a precise range of frequencies, all with the same gain (amplitude of filtering). These functions are unrealistic as we cannot have gain without phase and the infinitely steep cut-off (the edges of the regions) cannot be achieved. The Fourier pair of a pulse function is a sinc function, so our filter imposes a sinc-shaped response in the time domain (remembering that multiplication in the frequency domain is convolution in the time domain, so the circuit output is its convolution with the signal to be filtered). As these are continuous signals, the filters are implemented via circuits and there is no circuit which gives a sinc-shaped response function (that was after all the motivation for the flat-top window) so we need an approximation. Fourier analysis allows us to understand the frequencies involved.

So why do we need these filters? Let us consider a speaker system, illustrated in Figure 2.22, where we have three units: one for bass (the sub-woofer for low frequencies), one for mid-range and one for treble (the tweeter for high frequencies). (This is for illustration, noting that the units can be integrated into a single one.) A speaker for lower frequencies needs to be physically larger than one for high frequencies, so that we can maintain the volume of the different frequencies without distortion. As the speakers are different units, we need a circuit to split the signal: that is the filter which is the crossover network used to select which high-frequency signals to send to the tweeter, which to the mid-range, and which low-frequency ones to the sub-woofer. The cut-off frequency for the low-pass filter is set at 300 Hz, suitable for a sub-woofer system, the mid-range is between around 2.3–3 kHz, and the tweeter from 3 kHz. We shall find a real filter that is used, later. First, we shall look at some basic filters.

FIGURE 2.22 Using a filter within a speaker system

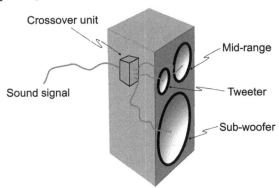

Crossover unit

Mid-range

Sound signal

Tweeter

Sub-woofer

2.6.1.2 RC networks and transfer functions: low-pass filters

Here we shall see how we can use the frequency domain to understand the filtering process and we shall use the convolution theorem to effect the analysis. We have actually met a filter already: the circuit in Figure 2.13 showed a low-pass filter which allows signals with low frequency to pass, but not signals with high frequencies. (More properly, high-frequency signals are attenuated – reduced – and those with lower frequency are not.) This can be understood by the frequency response of a capacitor which is an open circuit to low-frequency signals (the low frequencies do not pass) and a short circuit to high-frequency signals (letting the high frequencies through). These are shown as equivalent circuits in Figure 2.23 and at low frequency the capacitor disappears (the consequence of an open circuit) whereas at high frequency it is replaced by a piece of wire (the short circuit).

The function of the combined resistor and capacitor is the low pass filter's *transfer function*. This is developed using *Ohm's law* (voltage across = current through × resistance between; $v = i \times R$) as the voltage across the resistor is

$$v_i - v_C = v_R = iR \tag{2.55}$$

As the current is proportional to the rate of change of voltage across the capacitor

$$i = C\frac{dv_C}{dt}$$

Thus

$$v_i - v_C = RC\frac{dv_C}{dt}$$

We now see the time constant mentioned earlier, *RC*. (The solution to this was earlier in Section 2.3.1; a simpler derivation of this is by signal

FIGURE 2.23 Equivalent circuits to RC network at different frequencies: (a) low frequency; (b) high frequency

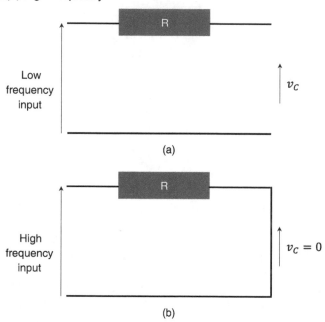

(a)

(b)

processing theory in Section 5.5.1.) The ratio of output to input voltage is the transfer function

$$\frac{v_C}{v_i} = \frac{1}{1 + RCs}$$

where s is the *Laplace transform* which can be considered equivalent to the derivative function $s = d/dt$ (happily driving a big truck through signal processing theory and differential equations, the Laplace transform can be found later in Chapter 5). The *frequency response* is its (imaginary) frequency part for $s = j\omega$.

$$\frac{v_C}{v_i}(j\omega) = \frac{1}{1 + j\omega RC} \tag{2.56}$$

The gain is then the magnitude

$$\left|\frac{v_C}{v_i}(\omega)\right| = \frac{1}{\sqrt{1 + (\omega RC)^2}}$$

In Equation 2.56 if we set $\omega = 0$ then $v_C/v_i(0) = 1$ so low frequencies are allowed to pass, as in Figure 2.23. Conversely if we set $\omega = \infty$, then $v_C/v_i(\infty) = 0$, so the high frequencies reduce to zero. The frequency response of a low-pass filter is the plot of the function in

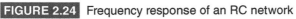

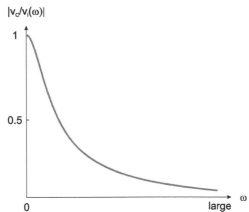

FIGURE 2.24 Frequency response of an RC network

Equation 2.56 given in Figure 2.24. By this it does not work very well: the pass region is rather short and the stop region for high frequencies is rather unclearly defined – it does, however, function correctly. The time constant defines the point where the low frequencies are passed and the high frequencies are stopped and this is known as the *cut-off frequency*. Fourier analysis allows us to determine the cut-off frequency, as it needs the frequency domain rather than the time domain. We shall now avoid the temptation to slip into signal processing stuff.

Let us see low-pass filtering in action, as illustrated in Figure 2.25. Here we have taken a signal and added some extra high frequencies to simulate noise. The low-frequency components are shown in Figure 2.25(b) and some high-frequency ones in Figure 2.25(c). When these are added together, we obtain a distorted signal, Figure 2.25(a). Now imagine that you are presented with the signal in Figure 2.25(a) and you know that we can filter it using an RC network. For a cut-off frequency of around 300 Hz we would need some values for the resistor and the capacitor: $R = 1k\Omega$ and $C = 0.5\mu F$ giving a time constant $RC = 5 \times 10^{-4}$. These are reasonably practical choices, for the same time constant one could equally choose $R = .1\Omega$ and $C = 5mF$, but the resistor would be a piece of wire and the capacitor enormous – or made from some rather expensive material. The output of the system is the result of convolution of the input with the transfer function as in Equation 2.36. To simplify calculation we apply the filter function of Equation 2.56 using the convolution theorem, and the whole Matlab instruction uses the FT of the input `fourier(p)` as

```
filtered_signal=ifourier(fourier(p)*1/(1+j*w*R*C));
```

The result is shown in Figure 2.25(d) and it is clearly close to the collection of low frequencies in Figure 2.25(b): we have achieved our

FIGURE 2.25 Applying an RC network low-pass filter to a signal: (a) original signal; (b) low-frequency components of (a); (c) high-frequency components of (a); (d) original signal low-pass filtered using an RC network

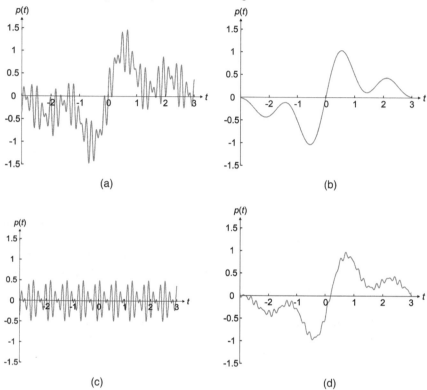

target and filtered out (most of) the high-frequency noise. Naturally, we could choose other values for the resistor and the capacitor because different values of the resistor and the capacitor, and thus the time constant, change the cut-off frequency of the low-pass filter.

2.6.1.3 CR networks and theory: high-pass filters

To achieve high-pass filtering, we simply swap the positions of the resistor and the capacitor to give the CR network in Figure 2.26.

In the *CR network* the current is again derived from voltage dropped across the capacitor, which is now the difference in voltage $v_C = v_i - v_R$. So

$$i = C\frac{d\left(v_i - v_R\right)}{dt}$$

FIGURE 2.26 CR network

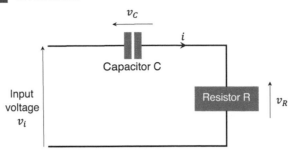

FIGURE 2.27 Frequency response of a CR network

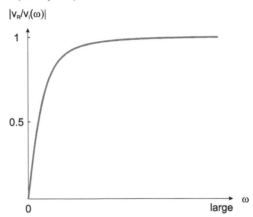

This forms the voltage across the resistor as

$$v_R = iR = RC\frac{d\,(v_i - v_R)}{dt}$$

By the Laplace transform

$$v_R = RCs(v_i - v_R)$$

Giving the transfer function

$$\frac{v_R}{v_i}(j\omega) = \frac{RCj\omega}{1 + RCj\omega} \tag{2.57}$$

By inspection, for high frequencies $v_R/v_i(\infty) = 1$ and for low frequencies $v_R/v_i(0) = 0$ (because the capacitor is equivalent to an open circuit) giving a high-pass filter with frequency response shown in Figure 2.27.

So let us use the convolution theorem to apply high-pass filtering to the signal of Figure 2.25(a), with result shown in Figure 2.28. We now

FIGURE 2.28 Applying the CR network high-pass filter

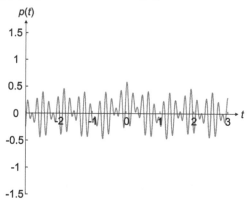

have the high frequencies which are close to those in Figure 2.25(c), the ones that were removed by the low-pass filter.

Evidently, the frequency response of the CR and RC networks is not particularly good for filtering. The phase of the CR and RC networks has not been shown and it is of similar form to the magnitude, so neither the gain nor the phase are linear or constant. Their virtue is simplicity because they are constructed from passive components only. (In VLSI you don't need capacitors: just leave a gap and that will give capacitance, and of a controlled amount too.) To improve matters we seek closer approximations to the functions in Figure 2.21. There are many of these, all with different advantages/disadvantages. We now consider the one that is often used in speaker crossover networks and stop there. Though we have just used the convolution theorem to apply the filters we shall use the frequency domain only for understanding in the next section.

2.6.2 Bessel filters

The circuits for active filters tend to include operational amplifiers, so we shall consider only the transfer functions and their frequency response. We shall look at one of the large selection, largely because of its association with hi-fi. The *Bessel filter* derives from a Bessel polynomial and a third order filter's transfer function *TB* is

$$TB(s) = \frac{15}{s^3 + 6s^2 + 15s + 15} \tag{2.58}$$

Again via $s = j\omega$ the gain of this circuit's frequency response is

$$G(\omega) = |TB(j\omega)| = \frac{15}{\sqrt{\omega^6 + 6\omega^4 + 45\omega^2 + 225}}$$

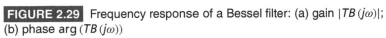

FIGURE 2.29 Frequency response of a Bessel filter: (a) gain $|TB\,(j\omega)|$; (b) phase arg $(TB\,(j\omega))$

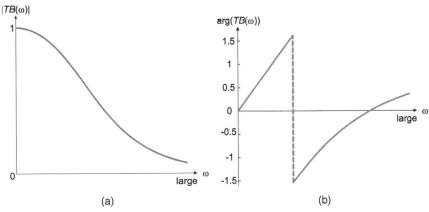

(a)   (b)

The phase is

$$\emptyset\,(\omega) = \arg\,(TB\,(j\omega)) = \tan^{-1}\left(\frac{15\omega - \omega^3}{15 - 6\omega^2}\right) \tag{2.60}$$

These are plotted in Figure 2.29 and show that the phase is linear when the magnitude is approximately constant. This is the 'you don't get anything for free' time, because magnitude and phase are intimately linked. It is important in audio that the phase is linear, otherwise some very fancy effects would ensue (wide sound – deep stereo imaging – introduced in the 1980s was largely a product of phase manipulations). Linearity is why the Bessel filter is used for crossover networks in audio circuits (though cheapskate producers might use RC and CR networks).

There are many more filters because the approximations are invariably a compromise between magnitude and phase. One example is *Tchebichev* filters which have a controlled ripple in the passband, but the phase performance is not as well controlled. We shall not go into any depth on these filters, because this book is d'ohing Fourier, not signal processing.

2.7 SUMMARY

This chapter developed and analysed the FT for continuous signals and sets down much of the theory we shall need later. The transform shows the frequency components of a signal and is bidirectional, allowing reconstruction via the inverse FT. If it wasn't bidirectional, it would simply be of naff all use. Some of the analysis will be used in later chapters, and some of it concerns basic functions that we shall meet not only in development of this material but also in its application. There are transform pairs for many functions, shown here for pulse and Gaussian functions. We can construct complex real-world signals by adding up lots of these functions, so their pairs show us nicely what's going on in the frequency domain. The theories also lead to properties which we need to understand so as to be able to understand the basic nature of the processes here. These include linearity (shown by superposition), invariance to shift in time, scaling and differentiation. The properties also lead to applications and we can use the Fourier transform not only to understand and analyse but also to allow for sophisticated application. This is where we ran into convolution and correlation, which are used to understand the basic response of systems, as well as modulation, which allows us to enjoy the music in our cars. There is a darker side too, because this material is complicated, and we find in applications that it is often not so simple as to calculate just an FT, but by its nature (and the signals themselves) we also need windowing functions. These lead straight to the next chapter, which is where we use the FT to analyse, process and understand signals that are stored in computers; let us move on to sampled signals, Chapter 3.

The Discrete Fourier Transform

The previous chapter concerned analysing the music that results when you play an instrument; this chapter is about processing the music that is stored in your computer. First, we need to consider the sampling process (partly because if you get it wrong you will ruin the signal you have sampled – some songs are immune, like Chris Rea songs, because they are already awful). After sampling, we shall move to processing the sampled signals. We shall later meet one of the best algorithms in the whole of Computer Science.

3.1 THE SAMPLING THEOREM

3.1.1 Sampling signals

Sampling concerns taking instantaneous values of a continuous signal, as opposed to windowing, which selects a section. Physically these samples are the outputs of an analogue to digital (A/D) converter and which are then stored in a digital (computer) system. These samples are taken regularly, at a chosen frequency. For the 1-D signals we have so far the samples represent, say, a music recording or the microphone in your mobile phone. The *samples* are the values of the signal at sampling instants. Intuitively, the sampling process is just a question of engineering – we acquire and store samples. In fact, it is a devious process as it is actually much more complicated than that, because by sampling incorrectly we can end up with the wrong signal. That means we need to do it right...read on!

The sampling process is illustrated in Figure 3.1 where Figure 3.1(a) concerns taking samples (x) at a high frequency: the spacing between samples, Δ_t, is small compared with the amount of change seen in the signal of which the samples are taken. Here, the samples are taken sufficiently fast to notice the dip in the sampled signal. Figure 3.1(b)

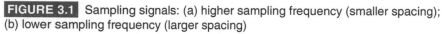

FIGURE 3.1 Sampling signals: (a) higher sampling frequency (smaller spacing); (b) lower sampling frequency (larger spacing)

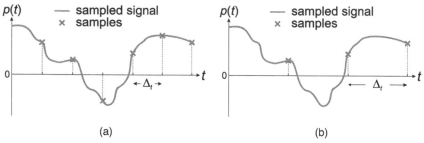

(a)   (b)

concerns taking samples at a lower frequency, compared with the rate of change of the sampled signal (its maximum frequency). Here, the dip in the sampled signal is not seen in the samples taken of it. The *sampling criterion* dictates the frequency at which the samples should be taken so that the dip is represented by the samples. How do we choose an optimum value for the sampling frequency or Δ_t? Before stating it, let us explore sampling a little more.

Let us consider taking the samples of a sine wave, as shown in Figure 3.2. Here, if we sample at a high frequency, Figure 3.2(a), the representation appears good and one can understand that the samples were taken from a continuous sine wave. Things become more difficult when we sample at a (much) lower frequency: in Figure 3.2(b) the samples appear to represent a sine wave of a different frequency – the samples do represent the signal from which they were derived, but the lower frequency signal appears more clearly. This is *aliasing*: if the sampling frequency is too low, the wrong signal appears (the alias of its original version).

We have not been able to find a human version of aliasing in sound signals (the change in frequency of a train's sound when it passes you is the Doppler Effect, not aliasing). The aliasing phenomenon actually occurs with human vision too and was understood by Dalí, as shown in Figure 3.3. Here, the original high-resolution image of his gorgeous painting in Figure 3.3(a) shows his wife's rear as she looks out to sea. The low resolution shows an alternative version. Those with poor eyesight can remove their glasses; for others, a lower-resolution version is given in Figure 3.3(b) – if you (as you are well advised to) visit the original in St Petersburg, Florida US, just step a distance away from it. The low-resolution image is actually a picture of Abraham Lincoln. Dalí knew this well; a small version of Lincoln can be seen just above Mrs. Dalí's left foot in Figure 3.3(a). The full title of the picture

> Dalí was profoundly uxorious, so Gala features quite a lot in his pictures. Female readers might prefer a man to be in the picture; men might prefer for Mrs Dalí to be looking the other way.

FIGURE 3.2 Aliasing by sampling at low frequency: (a) original continuous signal, frequency 1 kHz; (b) signal sampled at 11.2 kHz; (c) signal sampled at 0.9 kHz

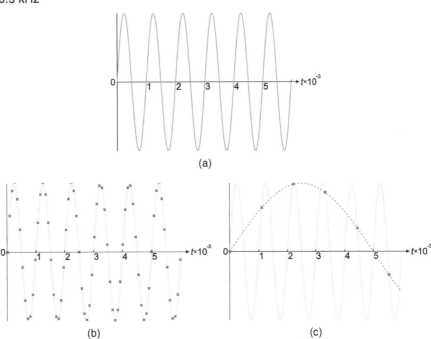

(a)

(b)   (c)

FIGURE 3.3 Aliasing in art [Dalí] © Salvador Dalí, Fundació Gala-Salvador Dalí, DACS 2021 (a) *Gala Contemplating the Mediterranean Sea*; (b) *Which at Twenty Meters Becomes the Portrait of Abraham Lincoln (Homage to Rothko)*

(a)   (b)

is subtitle (a) + subtitle (b), which lacks finesse. Another example visual sampling illusion is shown for images in the next Chapter.

3.1.2 Sampling process in the frequency domain

Figure 3.4 shows a sampling frequency and its frequency spectrum. In Figure 3.4(a) Δ_t is small so its reciprocal in Figure 3.4(b) is large – as in the relationship between time and frequency. So what's the explanation for this complex pair eh? Well, it's a bit of a stretch... From Wikipedia the function 'allows modelling sampling by multiplication with it, but it also allows modelling periodisation by convolution'. Let's have a look, with shorter words. The functions in Figure 3.4 are sets of delta functions known as a *Dirac comb*, because – well – they look like one! It is also referred to as the *sha* function because it looks like the Cyrillic letter Ш.

So how does this relate to the signals we have? As we see in Figure 3.5, we have a signal that is being sampled in Figure 3.5(a) and the spectrum of the signal before it was sampled is in Figure 3.5(b). The spectrum exists between the frequencies $-f_{max}$ and f_{max}. By the convolution theorem, convolution in the time domain is multiplication in the frequency domain: to sample a signal we multiply it with the function in Figure 3.4(a), and this implies that we convolve the transform in Figure 3.5(b) with the transform of the sampling function in Figure 3.4(b). This means that the spectrum repeats in the frequency domain (we said sampling was devious eh!). This repetition is by virtue of the sampling process and nothing else. The time-domain data is sampled, and so the frequencies are sampled too. Alternatively, the transform pair of a sampling function is another sampling function. We are only interested in the frequencies between $-f_{max}$ and f_{max} (the shaded section) so we can apply a windowing function (or a low-pass filter) to retrieve the frequencies in the range of interest. We are using frequency f for convenience, and the discussion could equally be phrased in terms of the angular frequency $\omega = 2\pi f$.

> A nit comb is one with a high sampling frequency? Ш is normally called sha though by Wikipedia it's 'the voiceless postalveolar fricative' (you might not need to know that!).

FIGURE 3.4 Sampling function transform pair: (a) sampling function; (b) Fourier transform of sampling function

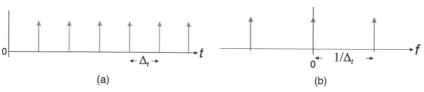

(a)

(b)

FIGURE 3.5 Sampling a signal: (a) sampled signal; (b) Fourier transform of signal before sampling; (c) Fourier transform of sampled signal

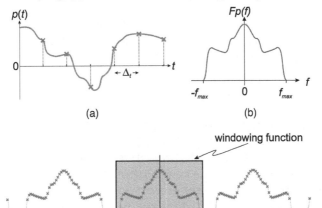

(a)

(b)

(c)

For the repetition in the FT relationship, the comb – or sampling – function is made up from a series of time-shifted delta functions

$$\text{comb}\,(t; \Delta_t) = \sum_{k=-\infty}^{\infty} (\text{shifted delta}) = \sum_{k=-\infty}^{\infty} \delta\,(t - k\Delta_t) \qquad (3.1)$$

and the *sifting* property of the delta function implies that information is only acquired at the sampling points and the intermediate values are not retained

$$\int_{-\infty}^{\infty} p\,(t)\,\delta\,(t - \Delta_t)\,dt = p\,(\Delta_t)$$

The transform of the comb function is

$$\mathcal{F}\,(\text{comb}) = \frac{2\pi}{\Delta_t} \sum_{k=-\infty}^{\infty} \delta\,(\omega - k\omega_s) \qquad (3.2)$$

which gives the mathematical relationship between Figure 3.4(a) and Figure 3.4(b). The inverse relationship between time and frequency maintains: the spacing is Δ_t in the time domain and $1/\Delta_t$ in the frequency domain. Given a continuous signal $p\,(t)$, its sampled version in Figure 3.5(a) is

$$\mathbf{p} = p\,(t)\,\text{comb}\,(t; \Delta_t) \qquad (3.3)$$

Via the sifting property of the delta function

$$\mathbf{p} = \sum_{k=-\infty}^{\infty} p\,(k\Delta_t)\,\delta\,(t - k\Delta_t)$$

Because

$$\mathcal{F}(\mathbf{p}) = \frac{1}{2\pi}\mathcal{F}(p(t)) * \mathcal{F}(\text{comb}(t; \Delta_t))$$

and the spectrum of the comb function repeats according to Equation 3.2 (as in Figure 3.4(b)), the signal's spectrum is repeated as in Figure 3.5(c) by virtue of the sampling process. So

$$\mathcal{F}(\mathbf{p}) = \frac{1}{\Delta_t}\sum_{k=-\infty}^{\infty} Fp(\omega - k\omega_s) \tag{3.4}$$

Thus the samples of the signal's spectrum in Figure 3.5(b) repeat at intervals of $f_s = 1/\Delta_t$, as in Figure 3.5(c).

The crux of the matter is the repetitive nature of the transform of the sampling function and the convolution of the two signals. A more intuitive explanation for the repetition derives from convolution. Earlier, we noted in Equation 2.11 that the FT of a delta function is unity. This is convolved with the FT of a pulse, Equation 2.5, where the pulse is short and zero for the rest of the sampling interval. We then multiply the two transforms together and the result is a series of sampling functions. Alternatively, for another intuitive explanation, let us consider that the Gaussian function (Figure 2.7) is actually made up of the summation of a set of closely spaced (and very thin) Gaussian functions. Then, because the spectrum for a delta function is infinite, as the Gaussian function is stretched in the time domain (eventually to be a set of pulses of uniform height), we obtain a set of pulses in the frequency domain, spaced by the reciprocal of the time-domain spacing.

Aliasing, by Figure 3.2, requires that we sample at the correct rate and we can determine this rate by the spectrum of the signal with different values of the sampling frequency f_{sample}. The process by which the correct rate is chosen is illustrated in Figure 3.6. Here, in Figure 3.6(a) we have the spectrum of a signal that has been sampled at an appropriate frequency, as in Figure 3.1(a). In Figure 3.6(c) we have the spectrum of a signal that has been sampled at too low a frequency, as in Figure 3.1(b). The minimum sampling frequency is that shown in Figure 3.6(b), and this is where the repeated spectra just touch. In Figure 3.6(c) the high frequencies are affected by those in the repeated spectrum – this is the aliased information. When the high frequencies are aliased, retrieval of the frequencies within the range of interest will recover high frequencies that have been corrupted. In Figure 3.6(a) the spectra do not intersect and the high frequencies are unaltered. Where the spectra touch in Figure 3.6(b), the sampling frequency is exactly twice the maximum frequency in the original signal

$$f_{\text{sample}} = 2f_{\text{max}} \tag{3.5}$$

This is the *Nyquist–Shannon sampling criterion*, key point 5.

FIGURE 3.6 Sampling process in the frequency domain: (a) sampling at high frequency; (b) sampling at the Nyquist frequency; (c) sampling at low frequency, aliasing the data

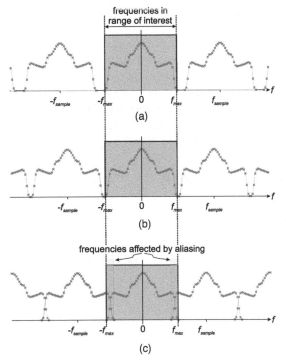

Key point 5

In order to be able to reconstruct a (discrete) signal from its samples we must sample at minimum at twice the maximum frequency in the original signal.

The use of the word 'reconstruct' is important: if the sampling is performed at an inappropriate rate, then we cannot return to where we started from. The *Whittaker–Shannon interpolation* formula or *sinc interpolation* describes how a function can be reconstructed from a set of samples. If the sampling frequency is correctly chosen, at or above the Nyquist frequency, then the reconstruction is perfect. Given the ubiquity of computers and stored music, speech, and so on, this is important stuff indeed. If we are sampling speech with a maximum frequency of 5 kHz, then we need to sample at minimum at 10 kHz. This implies that we need to sample a castrato singer more frequently than other people (and that sounds like a rather dicey proposition indeed). So we can now sample signals to form their discrete version and we will sample at minimum at the Nyquist rate. There are *anti-aliasing* filters that reduce a signal's bandwidth to a Nyquist rate and these can be analog (before sampling) like Figure 2.13 for audio. A higher inter-mediate sampling frequency can give *oversampling,* which leads to the better sound reproduction of the higher frequencies. That is more about audio and we can now move to the discrete Fourier transform (DFT). The need to represent data efficiently has more recently spread to discrete-time signals defined over irregular domains that can be described by graphs, which are sets of connections and vertices [Tanaka et al., 2020]. These graphs describe data other than regularly sampled speech signals and include architectures such as neuronal and social networks. This mandates an alternative approach say to selecting ver-tices and adapting to local structure, and will doubtless develop more as machine learning increases to be of importance.

3.2 THE DISCRETE FOURIER TRANSFORM

3.2.1 Basic DFT

The *DFT* of a set of N points p_x (ideally sampled at a frequency, which at minimum equals the Nyquist sampling rate) into each sampled fre-quency Fp_u is

$$Fp_u = \frac{1}{N} \sum_{x=0}^{N-1} p_x e^{-j\frac{2\pi}{N}ux} \qquad u = 0, 1 \ldots N-1 \qquad (3.6)$$

where the scaling coefficient $1/N$ ensures that the transform's d.c. coefficient Fp_0 is the average of all samples because for $u = 0$ we have

$$Fp_0 = \frac{1}{N} \sum_{x=0}^{N-1} p_x$$

Equation 3.6 is the definition we shall use from now on though there are other definitions as discussed in Section 7.1.1. Equation 3.6 is a discrete analogue of the continuous FT: the continuous signal is replaced by a set of samples, the continuous frequencies by sampled ones and the integral is replaced by a summation. Again, the process is complex so the transform components have magnitude and phase and the transform is often plotted as magnitude. The process is shown in Figure 3.7 where we have samples of a pulse plotted (Figure 3.7(a)). Here there are 128 samples of the pulse for which most are zero and for 16 of them (starting at sample 56) $p_x = 1.0$. The magnitude of the DFT when plotted against the samples of frequency is given in Figure 3.7(b) and these points can be seen to be arranged in a shape (the sinc function) similar to the curve of the continuous FT of a pulse shown in Figure 2.2(b). The phase, Figure 3.7(c), is quite frankly a mess (by the wraparound of the atan function), which is perhaps why many

FIGURE 3.7 Sampled pulse and its discrete Fourier transform (DFT):
(a) sampled pulse; (b) magnitude of DFT of sampled pulse; (c) phase of DFT of
sampled pulse

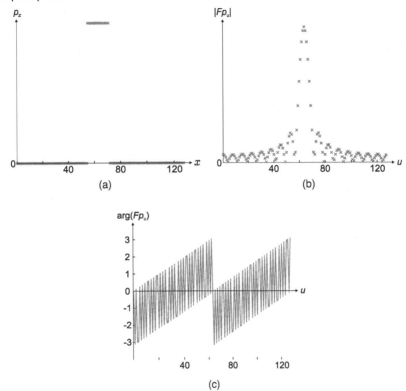

(a)

(b)

(c)

books do not show it. It is shown as a continuous plot rather than a
sampled one to reduce the mess (tidy your room lad!).

To understand how the DFT equation operates, we shall consider
the DFT of four sample points

$$Fp_u = \frac{1}{4} \sum_{x=0}^{3} p_x e^{-j\frac{2\pi}{4}ux} \qquad u = 0, 1, 2, 3 \qquad (3.7)$$

Expanding the equation we have

$$Fp_o = \frac{1}{4}\left(p_0 e^{-j\frac{2\pi}{4}0} + p_1 e^{-j\frac{2\pi}{4}0} + p_2 e^{-j\frac{2\pi}{4}0} + p_3 e^{-j\frac{2\pi}{4}0} \right)$$

$$Fp_1 = \frac{1}{4}\left(p_0 e^{-j\frac{2\pi}{4}1\times0} + p_1 e^{-j\frac{2\pi}{4}1\times1} + p_2 e^{-j\frac{2\pi}{4}1\times2} + p_3 e^{-j\frac{2\pi}{4}1\times3} \right)$$

$$Fp_2 = \frac{1}{4}\left(p_0 e^{-j\frac{2\pi}{4}2\times0} + p_1 e^{-j\frac{2\pi}{4}2\times1} + p_2 e^{-j\frac{2\pi}{4}2\times2} + p_3 e^{-j\frac{2\pi}{4}2\times3}\right)$$

$$Fp_3 = \frac{1}{4}\left(p_0 e^{-j\frac{2\pi}{4}3\times0} + p_1 e^{-j\frac{2\pi}{4}3\times1} + p_2 e^{-j\frac{2\pi}{4}3\times2} + p_3 e^{-j\frac{2\pi}{4}3\times3}\right)$$

Working through the integer parts (they are integer multiples of π), in *matrix form*, we have

$$\mathbf{Fp} = \frac{1}{4}\begin{bmatrix} 1 & 1 & 1 & 1 \\ 1 & -j & -1 & j \\ 1 & -1 & 1 & -1 \\ 1 & j & -1 & -j \end{bmatrix}\mathbf{p} \qquad (3.8)$$

We have done this to determine the matrix form of the eight-point DFT, which shows the structure more clearly than the four-point DFT. When we do the same for eight points, with .7 representing $\cos(\pi/4) = 1/\sqrt{2}$ so $e^{-j\frac{\pi}{4}} = \cos(\pi/4) - j\sin(\pi/4) = 1/\sqrt{2} - j/\sqrt{2}$ is written $.7 - .7j$. Thus we have

$$\mathbf{Fp} = \frac{1}{8}\begin{bmatrix} 1 & 1 & 1 & 1 & 1 & 1 & 1 & 1 \\ 1 & .7-.7j & -j & -.7-.7j & -1 & -.7+.7j & j & .7+.7j \\ 1 & -j & -1 & j & 1 & -j & -1 & j \\ 1 & -.7-.7j & j & .7-.7j & -1 & .7+.7j & -j & -.7+.7j \\ 1 & -1 & 1 & -1 & 1 & -1 & 1 & -1 \\ 1 & -.7+.7j & -j & .7+.7j & -1 & .7-.7j & j & -.7-.7j \\ 1 & j & -1 & -j & 1 & j & -1 & -j \\ 1 & .7+.7j & j & -.7+.7j & -1 & -.7-.7j & -j & .7-.7j \end{bmatrix}\mathbf{p}$$

$$(3.9)$$

The real parts of the matrix elements are shown row by row, for each frequency component, in Figure 3.8. Here the points are encircled and their horizontal position is indicated by a dotted line, with the underlying cosine wave shown as a solid line. For each frequency component n there are n full cycles of the cosine wave and the elements of the matrix exist at the different time instants. Since each element of each row scales a respective datapoint, we are fitting frequencies to the data, from low to high frequency. We could plot similar functions for the imaginary parts (the sine waves in the complex exponent) and it shows the same frequency structure. The DFT then determines how much of each of these frequencies is contained in the data, and that amount is the magnitude of each Fourier component. As with the continuous FT we are correlating sinusoids with the data. Note that two sine waves of different frequencies fit the same data for Fp_1 and Fp_7 (and for Fp_2 and Fp_6, together with Fp_3 and Fp_5). This reveals the *symmetry* intrinsic to the DFT, which we shall explore in Section 3.3.6.

FIGURE 3.8 Real parts of frequency components: (a) Fp, $\Re\left(e^{-j\frac{2\pi}{8}0}\right)$; (b) Fp_1, $\Re\left(e^{-j\frac{2\pi}{8}x}\right)$; (c) Fp_2, $\Re\left(e^{-j\frac{2\pi}{8}2x}\right)$; (d) Fp_3, $\Re\left(e^{-j\frac{2\pi}{8}3x}\right)$; (e) Fp_4, $\Re\left(e^{-j\frac{2\pi}{8}4x}\right)$; (f) Fp_5, $\Re\left(e^{-j\frac{2\pi}{8}5x}\right)$; (g) Fp_6, $\Re\left(e^{-j\frac{2\pi}{8}6x}\right)$; (h) Fp_7, $\Re\left(e^{-j\frac{2\pi}{8}7x}\right)$

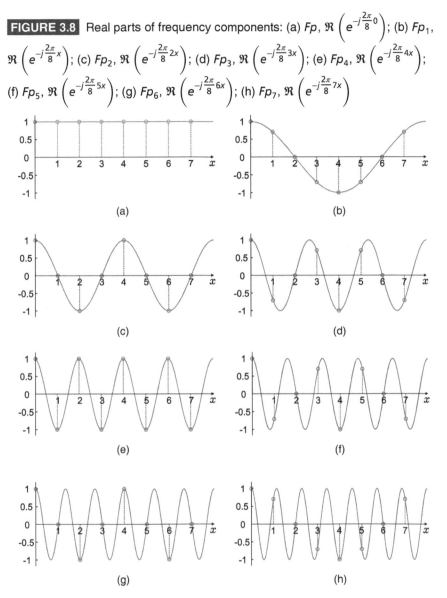

3.2.2 Inverse DFT

The FT would be of little use if there was no inverse: that is what leads to the variety of understanding, applications and derivatives. The *inverse DFT* is given by

$$p_x = \sum_{u=0}^{N-1} Fp_u e^{j\frac{2\pi}{N}ux} \qquad x = 0, 1 \ldots N - 1 \qquad (3.10)$$

Like the continuous inverse FT this has a change from the forward transform in the polarity of the exponential and in the scaling function. Let us consider reconstructing a sampled version of the continuous signal. The sampled signal and its DFT are shown in Figure 3.9. In the frequency domain, most of the sampled frequencies are zero, and only a few have some amplitude (similar to those in Figure 2.1, as expected).

There are of course arithmetic errors in the inverse DFT as there are in any computer-based calculation: numbers are not represented by an infinite number of bits. Figure 3.10(a) shows the result of the inverse DFT applied to the transform whose magnitude is shown in Figure 3.9(b). Clearly, Figure 3.10(a) shows a signal that appears to be the same as the signal in Figure 3.9(a), but there are small errors in the computer arithmetic: the magnitudes of the differences between the points in Figure 3.9(a) and those in Figure 3.10(a) are shown in Figure 3.10(b). At 10^{-14} these are very small indeed (and imperceptible), but they do exist. We shall return to this later. If the signal was aliased, these error values would be much larger.

So let us see the contributions of various components of the inverse DFT and the reconstruction error. This is shown in Figure 3.11 where Figure 3.11(a) is the reconstruction by the first two components (the d.c. component is zero so this is just adding the first frequency component), and Figure 3.11(d) shows the error in that reconstruction – which is considerable. The approximation is closer in Figure 3.11(b) with less error, Figure 3.11(e). Reconstruction by four components leaves out one component as in Figure 3.11(c) and that is the error component shown in Figure 3.11(f). As we increase the number of components, the signal increasingly approaches its original version, and the error

FIGURE 3.9 An arbitrary signal and the magnitude of its discrete Fourier transform (DFT): (a) sampled signal; (b) |DFT of sampled signal|

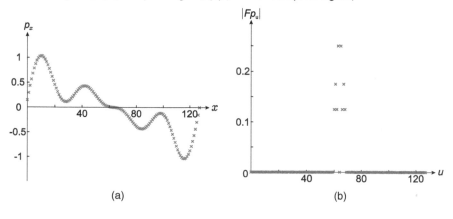

FIGURE 3.10 Inverse discrete Fourier transform (DFT) and reconstruction error: (a) inverse DFT applied to data in Figure 3.9(b); (b) error: result of |(a) − Figure 3.9(a)|

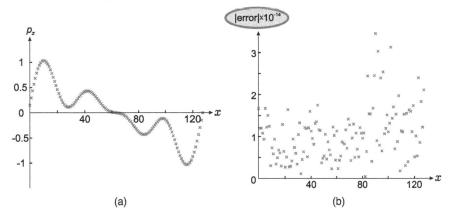

(a)　　　　　　　　　　　　　　　　(b)

decreases. For five components (or more) the reconstruction is precise, as in Figure 3.10(a) (with the arithmetic error shown in Figure 3.10(b)). The original signal is

$$p_x = 0.5\sin(x \times 2\pi/N) + 0.25\sin(2x \times 2\pi/N)$$
$$+ 0.35\sin(3x \times 2\pi/N) + 0.25\sin(4x \times 2\pi/N)$$

so Figure 3.11(a) is the first frequency component $0.5\sin(x \times 2\pi/N)$ and the error in Figure 3.11(f) is the highest frequency component $0.25\sin(4x \times 2\pi/N)$.

It is worth reiterating the points concerning reconstruction and the sampling theorem in Section 3.1.2. If a signal is sampled at an appropriate frequency, then the reconstruction is perfect, as it is here.

3.2.3 Visualising the DFT data

There are (quite) a few ramifications to the process. That's partly why we wrote this book! We shall do some of them right now, and get to the others soon. First off, according to sampling analysis, the spectrum repeats to infinity. This is *replication* and it can be shown analytically. We are concerned with what exists beyond the original N samples, so we shall be concerned with the DFT for a sample index $u + mN$, where m is an integer. So

$$Fp_{u+mN} = \frac{1}{N}\sum_{x=0}^{N-1} p_x e^{-j\frac{2\pi}{N}x(u+mN)}$$

$$= \frac{1}{N}\sum_{x=0}^{N-1} p_x e^{-j\frac{2\pi}{N}xu} \times e^{-j2\pi xm}$$

FIGURE 3.11 Reconstruction by the inverse discrete Fourier transform (DFT) and its error: (a) summation of first two DFT components; (b) summation of first three DFT components; (c) summation of first four DFT components; (d) error in (a); (e) error in (b); (f) error in (c)

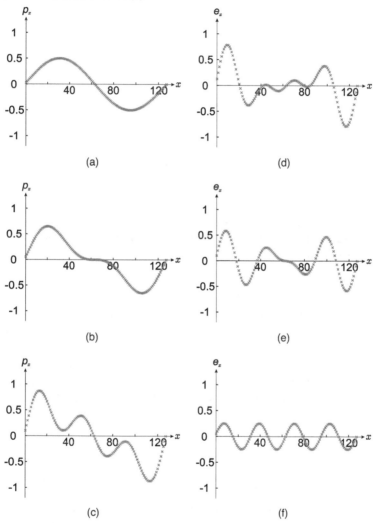

If the product xm is an integer, then $e^{-j2\pi xm}$ is formed from an integer multiple of 2π. Thus, for an arbitrary integer n, $e^{-jn2\pi} = \cos n2\pi - \sin n2\pi = 1$. As such, the FT repeats every N samples

$$Fp_{u+mN} = Fp_u \qquad (3.11)$$

By implication, if the spectra repeat indefinitely, then so does the signal in the time domain. Of course, this does not happen in reality and is an assumption implicit in Fourier theory. The consequences of this assumption will appear in the results provided by Fourier analysis, as we have just seen, though not when the assumption is anticipated. Let us summarise in key point 6.

<blockquote>
Key point 6
</blockquote>

The transform pair of the sampling function implies that the spectrum of a sampled signal repeats indefinitely in the frequency domain.

As in Figure 3.6, the full spectrum of the sampled pulse given in Figure 3.12 repeats every N sample. We are only interested in the $N = 128$ samples shown within the window. When you first apply the DFT to raw data, you will see the points that are within the window in Figure 3.12(a). It can be difficult for students to reconcile this with the transforms shown in books, like Figure 3.7(b). To derive the conventional depiction of the transform, one has to shift the window of interest to the position shown in Figure 3.12(b). Now the low frequencies are near the centre, and the higher frequencies are the most distant from the centre. Part of the problem is that, in theory, negative time and negative frequency are possible (and necessary). When implementing the DFT in a computer, we can only have samples stored with positive time indices. So we shift the window when displaying or visualising the transform data so that it fits with conventional interpretation. When processing, there is no need for this shift, so long as

FIGURE 3.12 Sampling windows of the discrete Fourier transform (DFT): (a) DFT of raw data; (b) conventional display window – DFT of processed data

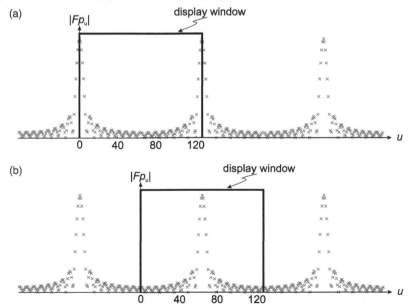

one remains aware of the way the transform is stored. In applications, for a transform of N data points, there is a maximum of $N/2$ frequency samples since when sampling at the Nyquist rate

$$f_{sample} = 2f_{max}$$

so by rearrangement

$$f_{max} = \frac{f_{sample}}{2} \tag{3.12}$$

There appear to be more frequencies associated with N points though some are repeated, as previously shown in Figure 3.8. As we shall find later, there is implicit symmetry in the DFT.

The magnitude of the transform is symmetric, and the position of the window is simply for visualisation. For the transform in Figure 3.7(b) there are 128 frequency components of interest, and these same points are those shown in Figures 3.12(a) and (b) where they repeat to infinity as in Equation 3.11. The unadulterated version of the DFT, Figure 3.12(a), actually goes from low to high frequency for the first 64 points, but then from high to low frequency in the next 64. This does not appeal to intuition, like Figure 3.12(b) (and as we saw earlier, Figure 3.7(b)). The windowing function here is for presentation only: a DFT is used to transform some data points and we seek to understand and use the result. When the DFT is used in its basic form we derive transform data like Figure 3.12(a) whereas it can be better understood by Figure 3.12(b).

It is convenient to do the visualisation automatically so that a conventional transform is presented. The process is actually quite simple and can be achieved just by multiplying each point p_x by -1^x before the transformation process. For the derivation of this, if $\cos(\pi) = -1$ then $-1 = e^{-j\pi}$ (the minus sign will keep the analysis neat), then the transform of the multiplied points is

$$\frac{1}{N} \sum_{x=0}^{N-1} p_x e^{-j\frac{2\pi}{N}xu} \times -1^x = \frac{1}{N} \sum_{x=0}^{N-1} p_x e^{-j\frac{2\pi}{N}xu} \times e^{-j\pi x}$$

$$= \frac{1}{N} \sum_{x=0}^{N-1} p_x e^{-j\frac{2\pi}{N}x\left(u+\frac{N}{2}\right)} \tag{3.13}$$

$$= Fp_{u+\frac{N}{2}}$$

which means that multiplying each point p_x by -1^x shifts the transform by half the number of points N. This shifts the display window from that in Figure 3.12(a) to the one shown in Figure 3.12(b) and we then automatically see the transform in its usual form (Figure 3.7(b)). (Matlab offers the function fftshift which by their documentation achieves the same positioning by swapping elements, which achieves the same result – and faster – but lacks insight into the operation of the DFT.) The visualisation process in Equation 3.13 is an example of how anticipating the basic nature of the FT can improve the analysis and its interpretation.

Let us see the transform of some real data. This is shown in Figure 3.13, where the music recording in Figure 3.13(a) is transformed to the version clearly showing its frequency content in Figure 3.13(c) – and some noise. As it is hard to see the signal in Figure 3.13(a), it has been expanded in Figure 3.13(b). It is actually the opening chord of the Beatles' A Hard Day's Night, about 2.5 s long – not to show the age of the author, more than the chord is incredibly identifiable. The spectrum of the chord in Figure 3.13(c) shows some clear peaks that could be the loudest notes when George Harrison played the chord on his lead guitar. There are other peaks and these are other frequencies, such as Paul McCartney's bass guitar and George Martin's piano. There is also noise, as that is everywhere, and there are other notes from John Lennon. The phase of the central portion of the spectrum is shown in Figure 3.13(d) (it would just be a mess like Figure 3.13(a) if the section was of longer duration). Because the d.c. coefficient of the spectrum can be very large, the spectrum can be plotted in logarithmic form as in Figure 3.13(e) (or even as power $20 \log_{10} |Fp_u|$ in dB). We shall only display the discrete points as an 'x' when looking at small sections of sampled signals, and otherwise the points are plotted as lines: these are the forms that will be used from now on. The chord is shown in Figure 3.13 (f) (and it would not be possible on a violin) and it doesn't show the silence that precedes the chord. There has actually been a short study of the chord using the FT [Brown, 2004] emphasising the importance of George Martin's piano, and this only became apparent when Fourier was deployed.

3.2.4   DFT in Matlab

For now, we are concerned with the implementation of the basic DFT. We shall later describe the FT that is usually used in Matlab (the fast Fourier transform (FFT) for those who have heard of it) in Section 3.6 and this gives Matlab's `fft` function, which is usually used to provide the DFT data. The basic DFT code here is the Matlab `function` in Code 3.1, where a vector of points is transformed into a vector of sampled frequencies stored in the vector `Fourier`. The summation in Equation 3.6 needs initialisation and then accumulation over the N points where $N = $ `cols`. The complex variable in Matlab is again `1j`. The result of that accumulation is the Fourier component at that frequency. Note that the vector indices in maths go from $0, 1, ..., N - 1$, whereas in Matlab they go from 1 to N (`1:cols`, in Matlab speak). This can give problems

FIGURE 3.13 Displaying the discrete Fourier transform of recorded music:
(a) a section of recorded music; (b) a closer look at the start of (a); (c) Fourier
spectrum; (d) phase of central portion; (e) Fourier spectrum in logarithmic form;
(f) chord.

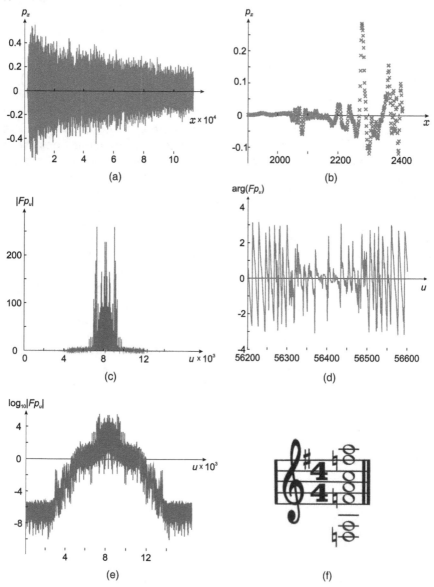

and care needs to be taken and it is why over loops u and x from 1
to *N* we have the factor *(u−1)*(x−1) to calculate the frequency correctly. In the inverse DFT the summation is over u discrete frequencies
and the multiplicand of the exponential is the Fourier component (and

the exponential is positive and the components are scaled by N, see the code at the book's website https://www.southampton.ac.uk/~msn/Doh_Fourier/). No one ever uses this implementation in applications as the FFT gives the same result and much faster, as we shall find.

Code 3.1 Discrete Fourier transform

```
function Fourier = DFT(vector)

%New vector is (basic) Discrete Fourier Transform
%(DFT) of input vector

vector=double(vector); %use double precision
[cols] = length(vector); %determine size of vector
Fourier(1:cols)=0; %largely to allocate storage
%we deploy Equation 3.6
for u=1:cols %for each frequency
  sumx=0;
  for x=1:cols %add up the transformation
    sumx=sumx+vector(x)*exp(-1j*2*pi*(u-1)*(x-1)/cols);
  end
 Fourier(u) = sumx/cols; %and now normalise it
end
```

Given that the code has two loops each covering 1 to N points, then the *computational cost* of the DFT is O (N^2), where O () denotes order and is largely a count of the number of operations. This is a rough measure because it misses the detail of the multiplications and additions. These calculations are performed in hardware for speed and multiplications tend to be slower than additions because the hardware is more complex. Subtraction uses (two's) complement arithmetic and has the same computational cost as addition, whereas division concerns multiplication by a reciprocal. Then there is the cost of the exponential function. O (N^2) is just an approximation, and it serves for later comparison. We shall discuss this again in Section 3.6.1 because speed is a critical issue limiting the general use of the basic DFT.

3.2.5 DFT pairs

A selection of DFT pairs is summarised in Table 8.5 for easy reference.

3.2.5.1 Pulse

In an analysis of a basic function (similar to that leading to Equation 2.5) we shall again analyse a pulse, but here a sampled one. (The argument remains that maths can handle simple signals, and in life, anything complex can be considered to be made up from lots of simple signals.) If the DFT is applied to the M samples of the pulse, which exists from sample 0 (p_0) to sample $M - 1$ (which is when the pulse ceases),

$$p_x = \begin{cases} A & x = 0, 1 \ldots M - 1 \\ 0 & \text{otherwise} \end{cases} \tag{3.14}$$

the DFT is given by

$$\mathcal{F}(\text{sampled pulse}) = Fp_u = \frac{1}{N} \sum_{x=0}^{M-1} A e^{-j\frac{2\pi}{N} ux} \tag{3.15}$$

Because the sum of a geometric progression can be evaluated according to

$$\sum_{k=0}^{n-1} a_0 r^k = \frac{a_0 (1 - r^n)}{1 - r} \tag{3.16}$$

the DFT of a sampled pulse is given by

$$Fp_u = \frac{A}{N} \left(\frac{1 - e^{-j\frac{2\pi}{N} uM}}{1 - e^{-j\frac{2\pi}{N} u}} \right) \tag{3.17}$$

By rearrangement, we obtain

$$Fp_u = \frac{A}{N} e^{-j\frac{\pi u}{N}(M-1)} \frac{\sin\left(\frac{\pi}{N} uM\right)}{\sin\left(\frac{\pi}{N} u\right)} \tag{3.18}$$

The modulus of the transform is

$$|Fp_u| = \frac{A}{N} \frac{\sin\left(\frac{\pi}{N} uM\right)}{\sin\left(\frac{\pi}{N} u\right)} \tag{3.19}$$

because the magnitude of the exponential function is 1. The magnitude $|Fp_u|$ is plotted for $M = 16$ and $N = 128$ in Figure 3.7(b) (and Figure 3.12). Compared with the FT of a continuous pulse, Equation 2.5, there are three components. One is the sinc function of the pulse, which is the numerator. This is scaled by the sinc function associated with the N point window, in the denominator; the last is the implicit sampling (shah) function. Other transform pairs are sampled versions of those shown previously in Section 2.1.5: a sampled sine wave is a Dirac function, and vice versa.

3.2.5.2 Gaussian

The next bit is a pain. However, the book aims to be complete, so we have to show that the DFT of a sampled version of a Gaussian is also a Gaussian. It is largely a repeat of Section 2.1.5, for discrete signals. For a sampled Gaussian function

$$g(x;\sigma) = g_x = \frac{1}{\sqrt{2\pi\sigma^2}}e^{\frac{-x^2}{2\sigma^2}} \tag{3.20}$$

and its DFT is

$$\mathcal{F}(\text{sampled Gaussian}) = \mathcal{F}(g(x;\sigma)) = \frac{1}{N}\sum_{x=0}^{N-1}g_x e^{-j\frac{2\pi}{N}xu} \tag{3.21}$$

Here we go... by substitution

$$= \frac{1}{N}\sum_{x=0}^{N-1}\frac{1}{\sqrt{2\pi\sigma^2}}e^{\frac{-x^2}{2\sigma^2}}e^{-j\frac{2\pi}{N}xu}$$

by manipulation

$$= \frac{1}{N}\frac{1}{\sqrt{2\pi\sigma^2}}\sum_{x=0}^{N-1}e^{-\left(x^2+\frac{j4\sigma^2\pi xu}{N}\right)/2\sigma^2}$$

and completing the square

$$= \frac{1}{N}\frac{1}{\sqrt{2\pi\sigma^2}}\sum_{x=0}^{N-1}e^{-\left(\left(x+\frac{j2\sigma^2\pi u}{N}\right)^2+4\sigma^4\left(\frac{\pi u}{N}\right)^2\right)/2\sigma^2}$$

by an ouch

$$= \frac{1}{N}e^{-\left(4\sigma^4\left(\frac{\pi u}{N}\right)^2\right)/2\sigma^2}\frac{1}{\sqrt{2\pi\sigma^2}}\sum_{x=0}^{N-1}e^{\frac{-\left(x+\frac{j2\sigma^2\pi u}{N}\right)^2}{2\sigma^2}}$$

by the area under a sampled Gaussian

$$= \frac{1}{N}e^{-\left(\frac{\sigma^2}{2}\left(\frac{2\pi u}{N}\right)^2\right)}\frac{1}{\sqrt{2\pi\sigma^2}}\sqrt{2\pi\sigma^2}$$

and thus we get

$$\mathcal{F}(g(x;\sigma)) = \frac{1}{N}e^{-\left(\frac{\sigma^2}{2}\left(\frac{2\pi u}{N}\right)^2\right)} \tag{3.22}$$

and thus, like Equation 2.14, the DFT of a Gaussian is another Gaussian. Was that really worth it? We shall investigate some properties now.

3.3 PROPERTIES OF THE DFT

A selection of DFT properties is summarised in Table 8.4, for easy reference.

3.3.1 Basic considerations

To analyse the properties of the DFT we need to change notation slightly. This is because we are dealing with sets of points, and we need a compact way to represent them. The notation that we shall use for a discrete signal consisting of N points is

$$\mathbf{p}[x] = p_x \quad x = 0 \ldots N - 1 \tag{3.23}$$

We have already analysed DFT replication previously in Section 3.2.3. We shall move on to the general properties of the DFT, which are similar to those of the continuous FT but are formed from sampled points, thus with different ramifications.

3.3.2 Linearity/Superposition

Superposition is also the same as in Section 2.2.1: both the continuous and discrete FTs are linear operations. As before, we need to show

$$\mathcal{F}(\text{signal } 1 + \text{signal } 2) = \mathcal{F}(\text{signal } 1) + \mathcal{F}(\text{signal } 2)$$

so

$$
\begin{aligned}
\mathcal{F}(\mathbf{p} + \mathbf{q}) &= \frac{1}{N} \sum_{x=0}^{N-1} (p_x + q_x)\, e^{-j\frac{2\pi}{N}ux} \\
&= \frac{1}{N} \sum_{x=0}^{N-1} p_x e^{-j\frac{2\pi}{N}ux} + \frac{1}{N} \sum_{x=0}^{N-1} q_x e^{-j\frac{2\pi}{N}ux} \\
&= \mathcal{F}(\mathbf{p}) + \mathcal{F}(\mathbf{q})
\end{aligned}
\tag{3.24}
$$

Job done! It is worth noting that there are nonlinear operations, which are those that do not obey superposition, say harmonic distortion in musical acoustics, and they can make life more complicated (especially with the DFT).

3.3.3 Time shift

Shift is the same as for the continuous FT, Section 2.2.2, in that the magnitude of the transform does not change with shift, but the phase does.

$$\mathcal{F}(\text{samples shifted by } \Delta) = \mathcal{F}(\mathbf{p}[x - \Delta]) = \mathbf{Fp}[u] \times e^{-j\omega\Delta} \tag{3.25}$$

Repeating Section 2.2.2 but for discrete signals we have

$$\mathcal{F}(\mathbf{p}[x - \Delta]) = \frac{1}{N} \sum_{x=0}^{N-1} p_{x-\Delta}\, e^{-j\frac{2\pi}{N}xu}$$

For $k = x - \Delta$ we have

$$= \frac{1}{N} \sum_{k=-\Delta}^{N-1-\Delta} p_k e^{-j\frac{2\pi}{N}(k+\Delta)u}$$

$$= \frac{1}{N} e^{-j\frac{2\pi}{N}\Delta u} \sum_{k=0}^{N-1} p_k e^{-j\frac{2\pi}{N}ku}$$

$$= e^{-j\frac{2\pi}{N}\Delta u} \mathcal{F}(\mathbf{p})$$

If $|e^{jx\text{any integer}}| = 1$, then the magnitude

$$|\mathcal{F}(\text{samples shifted by } \Delta)| = \left| e^{-j\frac{2\pi}{N}\Delta u} \right| |\mathcal{F}(\mathbf{p})| = |\mathcal{F}(\mathbf{p})| \qquad (3.26)$$

and the phase is affected by the term $e^{-j\frac{2\pi}{N}\Delta u}$.

3.3.4 Time scaling

In continuous time, the scaling was arbitrary. In *discrete time scaling*, the scale must be an integer factor since we are dealing with sampled signals. For an integer k

$$\mathbf{p}[x; k] = \begin{cases} \mathbf{p}[x/k] & \text{if } x \text{ is a multiple of } k; \ x = km \\ 0 & \text{otherwise} \end{cases} \qquad (3.27)$$

and because the reciprocal of time is frequency, the FT is

$$\mathcal{F}(\mathbf{p}[x; k]) = \mathbf{Fp}[ku] \qquad (3.28)$$

Note that this material might appear routine, but it remains of current interest because of reasons of implementation. One relatively recent study noted that their approach 'explicitly states the time/frequency axis scaling in terms of the multiplicative inverse of the scaling factor' [Talwalkar & Marple, 2010].

> Rama Chellappa once joked that in computer vision it's not that interest wanes when a problem appears to be solved, it's just that researchers just get bored with it.

3.3.5 Parseval's theorem (Rayleigh's theorem)

In the discrete case, *Parseval's theorem* relates the energy in the discrete components as

$$\sum_{x=0}^{N-1} |p_x{}^2| = k \sum_{u=0}^{N-1} |Fp_u{}^2| \qquad (3.29)$$

We shall avoid demonstrating this (and thus clarifying the constant k) as we do not need it in detail later on in this text (the demonstration is similar in nature to that for the continuous FT in Section 2.2.4). As

expected, the energy continues to be the same in the time and frequency domains. If it wasn't, we would be in trouble (or the first law
of thermodynamics would be past its sell-by-date – which it doesn't
have).

3.3.6 Symmetry

As previously shown in Section 3.2.1 there is implicit *symmetry* within
the DFT for real-valued signals (which we shall be using later). This
applies to the majority of sampled signals, where we assume the imaginary parts are zero. This is similar in notion to the symmetry of the
continuous transform but differs in the discrete nature. Noting that
the N points are indexed from 0 to $N - 1$, the frequency component
$N - u$ is

$$Fp_{N-u} = \frac{1}{N} \sum_{x=0}^{N-1} p_x e^{-j\frac{2\pi}{N}x(N-u)} = \frac{1}{N} \sum_{x=0}^{N-1} p_x e^{-j\frac{2\pi}{N}xN} e^{j\frac{2\pi}{N}xu} \tag{3.30}$$

Because

$$e^{-j\frac{2\pi}{N}xN} = e^{-j2\pi x} = 1 \qquad Fp_{N-u} = \frac{1}{N} \sum_{x=0}^{N-1} p_x e^{j\frac{2\pi}{N}xu} \tag{3.31}$$

As the sign of the exponent has changed, the component Fp_{N-u} is thus
the complex conjugate of the component Fp_u. So

$$\mathrm{Re}\left(Fp_{N-u}\right) = \mathrm{Re}\left(Fp_u\right) \tag{3.32}$$

$$\mathrm{Im}\left(Fp_{N-u}\right) = -\mathrm{Im}\left(Fp_u\right) \tag{3.33}$$

This is why for the eight-point transform in
Section 3.2.1 Fp_1 and Fp_7 had two cosine waves
at different frequencies matching the same
data. The relationship exposed in Equations
3.32 and 3.33 can also be seen in the matrix
form of Equation 3.9, say for Fp_3 and Fp_5 (rows
4 and 6). As the magnitude of the data is the

> Why is symmetry not a palindrome? With apology to etymologists, perhaps it should be 'symetemys'.

same, the magnitude of the Fourier components must be the same.
The magnitude shows the symmetry can be observed as

$$|Fp_{N-u}| = |Fp_u| \tag{3.34}$$

and this can be seen in Figure 3.12.

3.3.7 Differentiation

There is no differentiation for discrete signals and the counterpart is
to difference adjacent samples.

$$\mathbf{p}' \equiv \mathbf{p}[m] - \mathbf{p}[m-1]$$

By Fourier analysis, the frequency domain analysis of differencing is

$$\mathcal{F}\left(\mathbf{p}\left[m\right] - \mathbf{p}\left[m - 1\right]\right) = \left(1 - e^{-j\omega}\right)\mathcal{F}\left(\mathbf{p}\right) \tag{3.35}$$

The derivation of this first uses superposition

$$\mathcal{F}\left(\mathbf{p}\left[m\right] - \mathbf{p}\left[m - 1\right]\right) = \mathcal{F}\left(\mathbf{p}\left[m\right]\right) - \mathcal{F}\left(\mathbf{p}\left[m - 1\right]\right)$$

And by the shift theorem $\qquad = \mathcal{F}\left(\mathbf{p}\left[m\right]\right) - e^{-j\omega}\mathcal{F}\left(\mathbf{p}\left[m\right]\right)$

Thus, $\qquad\qquad\qquad\qquad = \left(1 - e^{-j\omega}\right)\mathcal{F}\left(\mathbf{p}\right)$

This is actually the domain of the z-transform and we shall cover that later in Section 5.6.

3.3.8 Importance of phase – DFT

We shall reprise our consideration of the importance of phase here for the DFT. We shall consider *gait biometrics*, which is recognising people by the way they walk. (OK, a lot of bias here as Mark and his students were the early pioneers of gait biometrics [Cunado et al., 1997; Nixon et al., 2010]; actually, Shakespeare started it all off since in *The Tempest* we can find 'Great Juno comes, I do know her by her gait' and seven other quotes in his plays that all imply gait is unique.) Gait is about swinging the legs and peoples' legs swing in different ways. We shall measure the inclination for two different subjects' thighs shown in Figures 3.14(a–c).

The front of the thigh's inclination is measured here as that of the line from the front of the top of the knee to the waistline (because these points are often clear, though can be obscured by a subject's swinging hand). The thigh's inclination over time, Figure 3.14(d), is a sinusoidal signal and the difference in shape from a sine wave is given by the muscles that pull the thigh up and then relax. There is some slight noise that can be observed on the measurements (gait is generally smooth, without abrupt changes). We shall measure the change in inclination over a gait period, which is the interval between heel strikes, which is when a heel first makes contact with the floor (and incidentally when shoelaces bounce up). An alternative interpretation is that this is when the two feet are furthest apart. When one foot is in contact with the floor, the other leg swings, and that is a step. A complete gait cycle comprises two steps, one of each leg. The variation of the inclination of the thigh is a periodic signal and the magnitude of the FT of the thigh's inclination for two different subjects is shown in Figure 3.14(e). There is little difference between the (windowed) magnitude spectra, as shown in Figure 3.14(f), because the basic mechanics of walking are largely the same for two subjects. What changes between subjects appears to be the timing, which is defined by the skeletomuscular structure and given by the phase of the FT. By plotting

FIGURE 3.14 Analysing human gait using the discrete Fourier transform (DFT): (a) subject 1 (at heel strike); (b) subject 1, later on; (c) subject 2 (at heel strike); (d) inclination of left thigh; (e) magnitude of DFT of inclination; (f) difference between magnitudes of DFT; (g) difference between phases of DFT

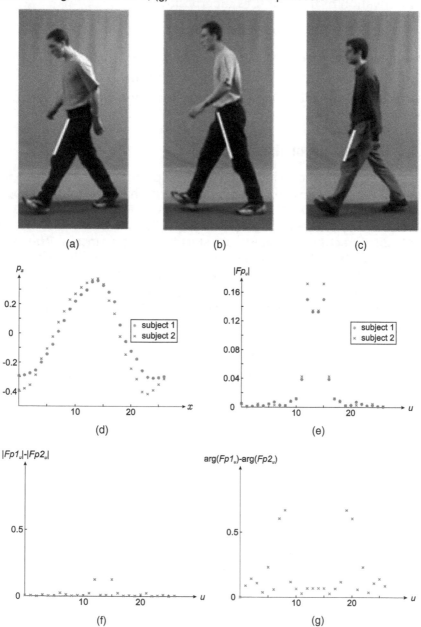

the difference of the phase spectra in Figure 3.14(g) the phase differs more for two people than the magnitude. (Note that figures (f) and (g) are normalised in amplitude to the maximum of either, for comparison, and are shown in absolute form so that order of analysis does not matter. There are other ways to normalise, depending on the intention of the graph, and the normalisation here is intended to show any difference between the relative influence of magnitude and phase.) The average normalised distance in phase is much larger than that of the magnitude (the average normalised phase is 0.17, which vastly exceeds 0.018 for the normalised magnitude) and the variance of the phase difference points is also much greater than the variance of the magnitude difference points. It is not that phase is more important than the magnitude; clearly, the phase cannot be ignored (and that is evident by its role in reconstruction by the inverse FT). In our early gait biometrics research we showed that the magnitude is important (since size generally is, points which are zero have no consequence) so people were recognised by unique descriptions that were derived from the product of the magnitude and the phase of the DFT of the inclination of their thighs [Cunado et al., 2003]. Gait has since been shown to be unique on very large databases of up to 100,000 subjects. Spot on Shakespeare!

3.3.9 Discrete data windowing functions

As seen earlier in Section 3.2.3, windowing functions are very important in visualising the FT. The importance is reflected in the nature of replication: since the transform repeats periodically, if the end points are not of the same value (the function is not periodic over the number of points), then there are discontinuities at the ends of the sampling window. The function is assumed (implicitly by sampling) to change very quickly at these points, as shown for a non-periodic function in Figure 3.15. So we use windowing to zap them (or more prosaically to conserve the energy of the high-order spectral density estimates).

  As an in-depth analysis of windowing was given in Section 2.5 for continuous signals, we shall just note the main discrete windowing functions here, in discrete form. There is naturally a discrete version of the Short-Time Fourier transform

$$STFp_u = \frac{1}{NW} \sum_{x=-\infty}^{\infty} p_x w(\tau - x) e^{-j\frac{2\pi}{NW}ux} \tag{3.36}$$

which performs the transform over shorter sections of data by using the window w to localise the transform to be computed over a specific section of NW data samples starting from time τ. When this is used

FIGURE 3.15 Replication in the discrete Fourier transform (DFT): (a) original signal **p**; (b) replication in DFT

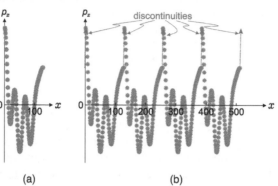

(a)   (b)

within an efficient means for calculating the whole FT there are natural considerations of spacing (the *hop size*) and the amount of *overlap* between the windowed sections of data.

The rectangular windowing function for sampled signals for *NW* points is denoted as

$$wR_x(NW) = \begin{cases} 1 & 0 \le x < NW - 1 \\ 0 & \text{otherwise} \end{cases} \tag{3.37}$$

and this introduces the sidelobes of the sinc function shown in Figure 3.7 so we use two functions that have a smoother transform. The discrete form of the Hann window is

$$wHAN_x(NW) = \begin{cases} 0.5 + 0.5\cos\left(\dfrac{\pi\left(x - \frac{NW}{2}\right)}{\frac{NW}{2}}\right) & 0 \le x < NW - 1 \\ \\ 0 & \text{otherwise} \end{cases} \tag{3.38}$$

The discrete form of the Hamming window is

$$wHAM_x(NW) = \begin{cases} 0.54 + 0.46\cos\left(\dfrac{\pi\left(x - \frac{NW}{2}\right)}{\frac{NW}{2}}\right) & 0 \le x < NW - 1 \\ \\ 0 & \text{otherwise} \end{cases}$$

$$\tag{3.39}$$

The discrete forms of Hanning and Hanning windows and their spectra are plotted in Figure 3.16 and these are of similar shape (and difference) shown for the continuous versions in Figure 2.18. There are of course other sampling functions, such as the flat-top window

FIGURE 3.16 Discrete windowing functions: (a) discrete Hanning and Hamming windows; (b) spectra of discrete windows

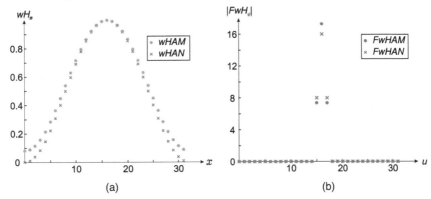

(a)   (b)

and there is excellent advice available [Harris, 1978], focused on estimation of harmonics in noise. A more recent approach to windowing with localised frequency analysis is wavelets, which are covered in Section 5.7.

3.4 DISCRETE CONVOLUTION AND CORRELATION

3.4.1 Discrete convolution

In general, *discrete convolution* for discrete signals (of unknown length) is defined as

$$\mathbf{p} * \mathbf{q} = \mathbf{p}[m] * \mathbf{q}[m] = \sum_{x=-\infty}^{\infty} \mathbf{p}[x] \times \mathbf{q}[m-x]$$

$$= \mathbf{q}[m] * \mathbf{p}[m] = \sum_{x=-\infty}^{\infty} \mathbf{q}[x] \times \mathbf{p}[m-x] \tag{3.40}$$

where the time-inversion process, which gives the memory function of Equation 2.33, is applied to one of the signals. Now this is a book on Fourier, so we are concerned with sampled signals and we shall have vectors of *N* points as in Equation 3.23 and by the replication theorem Equation 3.11 the Fourier spectrum repeats infinitely. By implication, if the implicit assumption is that (sampled) spectra repeat indefinitely, then the signal is assumed to repeat in the time domain. Discrete convolution for two sampled (periodic) signals each with *N* points is *circular convolution*

$$\mathbf{c\_conv} = \mathbf{p}[m] * \mathbf{q}[m] = \sum_{x=0}^{N-1} \mathbf{p}[x] \times \mathbf{q}[m-x] \quad m = 0, 1 \ldots N-1 \tag{3.41}$$

This is the conventional notation, but it rather obscures the circularity which arises from the use of a single period only as

$$c\_conv_m = \sum_{x=0}^{N-1} p_x \times q_{\mathrm{mod}(m-x,N-1)} \quad m = 0, 1 \ldots N-1 \qquad (3.42)$$

where mod is the *modulo* operator (the integer remainder of $(m-x)/(N-1)$). The mod function can be applied to either signal as in Equation 3.40 but not to both because convolution remains commutative as in Equation 2.39:

$$\mathbf{p} * \mathbf{q} = \mathbf{q} * \mathbf{p} \qquad (3.43)$$

As with the continuous version, it is also distributive and associative.

As with continuous convolution, discrete convolution evaluates the amount of overlap between one set of points, and another when it is time reversed. The overlap or summation is stored for each point. The notation can clearly be seen in the implementation of discrete convolution (which is circular convolution for periodic signals) in Code 3.2, which implements Equation 3.41. In Code 3.2 new_mod is a modulus operator returning a value in the range 1:cols, consistent with Matlab's vector indexing in circular convolution. The default form of Matlab's own convolution operator conv implements the general form of discrete convolution, Equation 3.40, where a signal potentially exists for all time. The default case of Matlab's conv operator then calculates the output for a longer period of twice the length, and care must be taken with its display. The operator c_conv, (Code 3.2) is circular for periodic signals only and closely follows Equation 3.42, so the output is the same length as the signals from which it was derived.

Code 3.2 Discrete circular convolution

```
function c_conv = discrete_conv(vector1,vector2)

cols=length(vector1); %determine size of vector
c_conv(1:cols)=0; %clear output storage
for m=1:cols
  c_conv(m)=0; %initialise output element
  for x=1:cols %and then add successive multiplications
    c_conv(m)=c_conv(m)+(vector1(x)*
            vector2(new_mod(m-x+1,cols)))); %Eq 3.41
  end
end
```

The use of Code 3.2 can be seen in Figure 3.18 where a pulse, Figure 3.18(a), is convolved with a delayed echo function, Figure 3.18(b). In convolution the response is reversed to form a memory function, Figure 3.18(c). If the input to the delayed echo system

FIGURE 3.17 Example digital filter

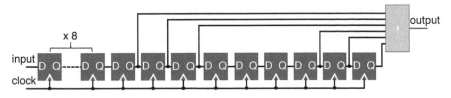

FIGURE 3.18 Illustrating discrete circular convolution: (a) pulse; (b) response; (c) memory function; (d) discrete convolution process and system output

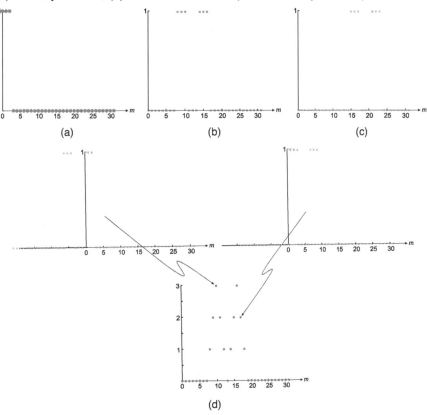

is a short pulse lasting for one sample only (a unit delta function), then the output will be three echoes, played twice, and occurring some time after the tit was tapped. A *digital filter* that could produce the signal is shown in Figure 3.17, which consists of delay elements followed by the addition of the delayed samples. The architecture is built around a shift register in which each D-type bistable stores its input (to become the Q output) at each clock edge. This can be implemented using a programmable gate array; digital hardware is much faster than using a computer, though it is much less general.

If the pulse is longer, what happens? Let us order a convolution takeout, and that will show us the result. The output and two stages of the process are shown in Figure 3.18(d) and we can see that the echoes add up to give a triangular waveform because they are adjacent. Early on, in the first stage shown three of the echo samples overlap with the pulse samples (so the pulse is echoed thrice, concurrently); in the second stage shown there are only two.

The convolution theorem also extends to discrete signals: convolution can use the FT, which is efficacious given the existence of the FFT. By the DFT the FT of two convolved signals, $*$, is

$$\mathcal{F}\left(\mathbf{p}[m] * \mathbf{q}[m]\right) = \mathcal{F}\left(\sum_{x=0}^{N-1} \mathbf{p}[x]\,\mathbf{q}[m-x]\right) \tag{3.44}$$

by substitution via Equation 3.6

$$= \frac{1}{N}\sum_{m=0}^{N-1}\left(\sum_{x=0}^{N-1} p_x q_{m-x}\right) e^{-j\frac{2\pi}{N}mu}$$

by reordering

$$= \frac{1}{N}\sum_{x=0}^{N-1} p_x \sum_{m=0}^{N-1} q_{m-x} e^{-j\frac{2\pi}{N}mu}$$

by Equation 3.25

$$= \frac{1}{N}\sum_{x=0}^{N-1} p_x \sum_{m=0}^{N-1} q_m e^{-j\frac{2\pi}{N}mu} e^{-j\frac{2\pi}{N}xu}$$

by grouping like terms

$$= \frac{1}{N}\sum_{x=0}^{N-1} p_x e^{-j\frac{2\pi}{N}xu} \sum_{m=0}^{N-1} q_m e^{-j\frac{2\pi}{N}mu}$$

and (by serendipity?)

$$= (\mathcal{F}\left(\mathbf{p}[m]\right) \times \mathcal{F}\left(\mathbf{q}[m]\right))_u$$

By this, the implementation of discrete convolution using the DFT is again achieved by multiplying the two transforms and differs from its continuous counterpart in the nature of the multiplication (products of vectors are either scalars or matrices, not vectors of points). For two sampled signals each with N points we have

$$\mathcal{F}\left(\mathbf{p} * \mathbf{q}\right) = \mathcal{F}(\mathbf{p}).\times \mathcal{F}\left(\mathbf{q}\right) \tag{3.45}$$

where $.\times$ denotes element-wise multiplication. For two vectors $\mathbf{a}$ and $\mathbf{b}$, each with N elements, the dot product is

$$\mathbf{a}.\times \mathbf{b} = \left[\, a_0 \times b_0 \ \ a_1 \times b_1 \ \ \ldots \ \ a_{N-1} \times b_{N-1} \,\right] \tag{3.46}$$

By definition, Equation 3.40, discrete convolution is commutative (Equation 3.43) and so order does not matter, the same for continuous convolution in Equation 2.39. This is illustrated in Figure 3.19 where all transforms are shown as magnitude and centred. Figure 3.19(a) is the FT of the pulse in Figure 3.18(a) and Figure 3.19(b) is the FT of the response in Figure 3.18(b). The product of the two transforms is

FIGURE 3.19 Discrete circular convolution via the Fourier transform:
(a) discrete Fourier transform (DFT) of pulse $|\mathcal{F}(\text{Figure 3.8(a)})|$; (b) DFT of
response $|\mathcal{F}(\text{Figure 3.8(b)})|$; (c) pointwise multiplication of transforms (a) × (b);
(d) inverse Fourier of pointwise multiplication $\mathcal{F}^{-1}(c)$

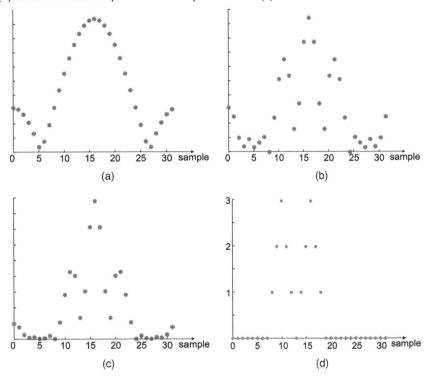

(a)

(b)

(c)

(d)

shown in Figure 3.19(c) and the result of the inverse FT is shown in
Figure 3.19(d).

We have so far considered circular convolution and not the *linear
convolution* of Equation 3.40. This is more the domain of systems analysis, and it is included for completeness. Because convolution concerns
inverting one signal and passing it over the other, the result of linear
convolution is twice as long as the duration of either of the signals.
This is illustrated in Figure 3.20 for a discrete version of the RC filter system response **q**, Figure 3.20(b), with a discrete pulse **p** as input,
Figure 3.20(a). Because convolution is commutative, Equation 3.43,
either signal can be reversed in time. The pulse input **p** is time reversed
(as in Section 2.3.1) and for computation the system response is *zero
padded*, Figures 3.20(c) and (d). In the zero padding process extra
points are added which are all set to zero, to enable the calculation process The position of the pulse is overlaid for sample 0 in
Figure 3.20(c) and for sample 15 in Figure 3.20 which is the maximum
intersection between the two signals, shown in Figure 3.20(e). Thus

FIGURE 3.20 Illustrating linear convolution: (a) sampled pulse; (b) samples of system response; (c) convolution process, sample 0; (d) convolution process, sample 15; (e) linear convolution result

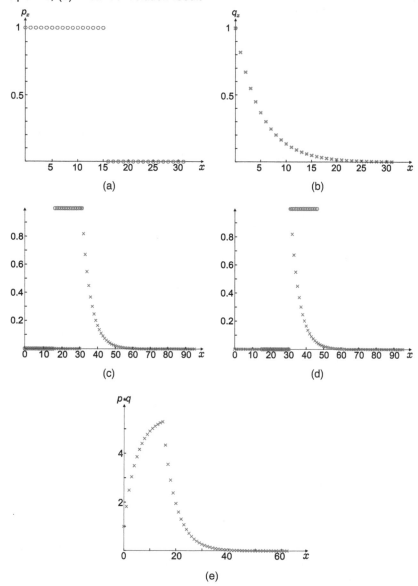

we obtain a discrete version of the signal in Figure 2.12, and it lasts for around twice the number of samples as in **p** or **q**, by virtue of the summation process. This is largely the same result as for Matlab's `conv` operator (though Matlab might use Fourier, which could have been done here too).

3.4.2 Discrete correlation

As with continuous correlation in Section 2.3.2, *discrete correlation* is given by

$$\mathbf{p}\,[m] \otimes \mathbf{q}\,[m] = \sum_{x=0}^{N-1} \mathbf{p}\,[x]\,\mathbf{q}\,[m+x] \qquad m = 0, 1 \ldots N - 1 \qquad (3.47)$$

and because the signals we are interested in here are periodic, this is *circular correlation*.

Discrete correlation is illustrated in Figure 3.21 aiming to identify the position of a triangular signal, Figure 3.21(a), first from a clean series of triangles (Figure 3.21(b)) and then from a noisy sequence (Figure 3.21(c)). The noisy version has been achieved by adding a mean-zero random function. The noisy sequence reflects corruption that can occur in a communications channel and the position of the triangles in the noisy signal is rather difficult to determine visually. The process of correlation is illustrated in Figure 3.21(d) (for the position of best match to the noisy signal where $m = 39$ which is one of the three best matches to the clean signal). Correlation (Figure 3.21(e)) shows three peaks, which are the positions of the (first points of the) triangles in Figure 3.21(b) as well as the effects of modulo computation (at the ends). Noise clearly affects the magnitude of the peaks in Figure 3.21(f) though it does not affect their positions. The result is actually very clear, despite the noise contaminating the original signal. This is because of the averaging process in correlation, which reduces the noise to its mean value. Should the positions of the three triangles be desired, then it is given by the positions of the three maxima. In this way, discrete correlation can be used to detect signals in (extremely) noisy data.

This can be implemented by the FT and by the previous section this is given by

$$\mathcal{F}\,(\mathbf{p} \otimes \mathbf{q}) = \overline{\mathcal{F}\,(\mathbf{p})} . \times \mathcal{F}\,(\mathbf{q}) \qquad (3.48)$$

where $\overline{\mathcal{F}\,(\mathbf{q})}$ denotes the complex conjugate, noting the symmetry relations in Equation 3.31, which gives

$$Fp_{N-u} = \overline{Fp_u} \qquad u = 0, 1 \ldots N - 1$$

As with continuous convolution, discrete convolution is also the result of inverting the time axis of one of the signals whose FTs are multiplied. So correlation can also be implemented in the frequency domain by inverting the time index of one of the signals. Here we shall invert $\mathbf{p}$ as

$$\mathcal{F}\,(\mathbf{p} \otimes \mathbf{q}) = \mathcal{F}\,((N-1) - \mathbf{p}) . \times \mathcal{F}\,(\mathbf{q}) \qquad (3.49)$$

FIGURE 3.21 Illustrating discrete correlation: (a) triangle p_x; (b) sequence of triangles q_x; (c) triangle sequence with added noise nq_x; (d) position of maximum overlap; (e) correlation with clean data $\mathbf{p} \otimes \mathbf{q}$; (f) correlation with noisy data $\mathbf{p} \otimes \mathbf{nq}$

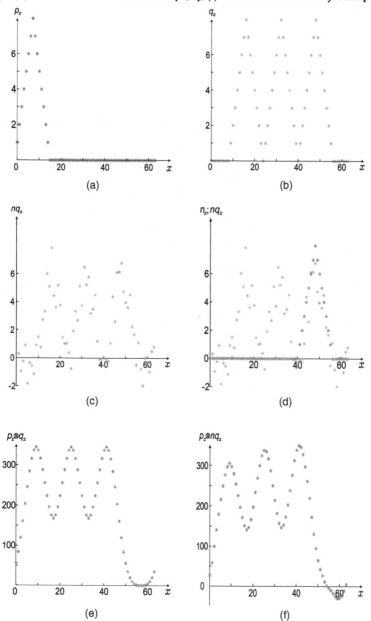

where the term $(N-1) - \mathbf{p}$ denotes inversion of the set of samples $\mathbf{p}$ along the time axis

$$\mathcal{F}((N-1) - \mathbf{p}) = Fp_{N-1-u} \qquad u = 0, 1 \dots N-1$$

Correlation is not commutative and order matters: the results depend on the order in which signals are presented. This implies that care must be taken in implementation: by Fourier, one gets a result dependent precisely on what has been specified (which is another confirmation of the need for reproducibility). As with the continuous version, correlation is distributive but not associative. Also, even functions are not affected by time reversal, so convolution and correlation can deliver the same results. When circular convolution is deployed, the window used to view signals can affect the result. One can use the centring process given in Equation 3.13 to correct for this, to ensure that the correct order of multiplication is retained. In this way the result of Fourier-based analysis depends largely on what has been done, as in any system. The advantage of Fourier is speed; the disadvantage is complexity.

We have so far considered circular correlation and not the *linear correlation*. As with convolution, this is more the domain of systems analysis, and it is included for completeness. Because correlation concerns passing one signal over the other and evaluating the match between the two, the result of linear correlation, like linear convolution, is almost twice as long as the duration of either of the signals. This is illustrated in Figure 3.22 for a target signal **q** in Figure 3.22(b), with a discrete ramp **p** as input in Figure 3.22(a). For computation, as in Section 2.3.2, the system response is zero padded, Figures 3.22(c) and (d). The position of the pulse is overlaid for the first intersection in Figure 3.22(c) and the second maximum intersection between the two signals in Figure 3.22(d). The match derived by correlation is shown in Figure 3.22(e). The match reflects the whole template and so is recorded at the position of the origin of the template. Thus we obtain a signal (with some similarity to continuous matching in Figure 2.14), which lasts for twice the number of samples as in **p** or **q** by virtue of the summation process. This is largely the same result as for Matlab's corr operator. Note that (Pearson's) *correlation coefficient* is a statistical measure of correlation between two variables, which is another measure of matching.

3.5 DIGITAL FILTERS; AVERAGING AND DIFFERENCING SAMPLES

To achieve low-pass filtering we seek to block the high frequencies, and conversely to block the low frequencies for high-pass filtering. For a low-pass filter, the difference between filtering and the windowing functions is that the symmetry implicit in the DFT, Equation 3.34, means that we have to block components either side of the d.c. coefficient. For a low-pass filter, which passes frequency components lower than

FIGURE 3.22 Illustrating linear correlation: (a) example signal; (b) samples of target signal; (c) correlation process, first intersection; (d) correlation process, second maximum; (e) linear correlation result $\mathbf{p} \otimes \mathbf{q}$

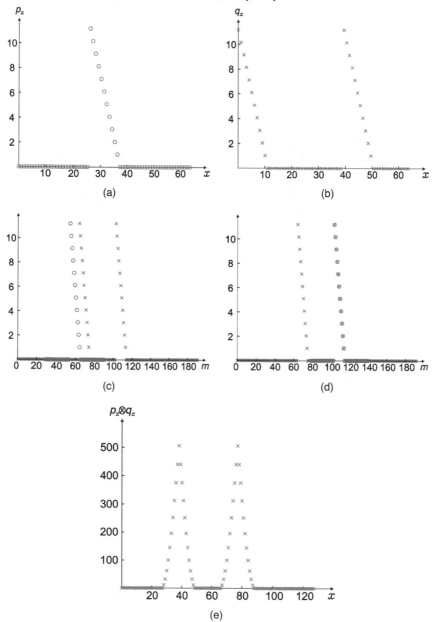

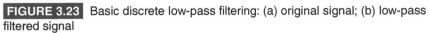

FIGURE 3.23 Basic discrete low-pass filtering: (a) original signal; (b) low-pass filtered signal

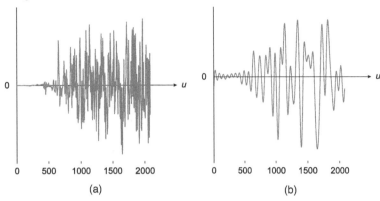

(a)   (b)

a chosen *cut-off* frequency, and a DFT, which has not been centred (so the low frequencies are at the beginning and end), we have

$$\mathbf{Lp} = low\_pass\left(\mathbf{Fp}; cut\_off\right) = \begin{cases} Fp_u & u < cut\_off \\ Fp_{N-u} & N - u > N - cutoff \\ 0 & otherwise \end{cases} \quad (3.50)$$

and the converse of this will achieve high-pass filtering. The application of Equation 3.50 to the Hard Day's Night chord with a cut-off frequency of 30 Hz is shown in Figure 3.23, which shows that the smoothing consistent with low-pass filtering has been achieved.

When you listen to the audio, this would certainly make Paul and Ringo cringe (or probably just throw up). Its main use is to filter noise, and of course there is little noise on a Beatles recording and we'd retain most frequency components, and not as few as 30. It is rather basic and by its equivalence to a pulse in the frequency domain, by the convolution theorem will have the effect of convolving a sinc function with the data, in the time domain. This is likely to be the reason for the variation at the start of the filtered signal which is not zero any more. There are much more sophisticated digital filters. The digital filter in Figure 3.17 incorporated delay, followed by addition. The addition of samples will average an output and this is low-pass filtering. A general *averaging filter* over A samples is

$$average_x = \begin{cases} \dfrac{1}{A} & 0 < x < A - 1 \\ 0 & otherwise \end{cases} \quad (3.51)$$

and when this is convolved with a signal, it will smooth it, though it is more usually performed as *direct averaging*

$$\text{average}(\mathbf{p}[m]; A) = \frac{1}{A} \sum_{x=0}^{A-1} \mathbf{p}[m] \qquad m = 0, 1 \ldots N - 1 \qquad (3.52)$$

Though this is appealingly simple, a quick look at the frequency domain properties reveals that it is not as attractive as it might seem. This is again because the averaging function is equivalent to a pulse and this has a transform pair, which is the sinc function, the same as in the low-pass filter of Equation 3.50. When we perform direct averaging of the form of Equation 3.52, or convolve the averaging function of Equation 3.51, the sidelobes of the sinc function will have a deleterious effect on the high-frequency content of the averaged signal. It is, therefore, propitious to perform *Gaussian averaging* where the template coefficients are given by

$$\text{Gaussian\_average}_x = \begin{cases} e^{\dfrac{-\left(x-\frac{A}{2}\right)^2}{2\sigma^2}} & 0 < x < A - 1 \\ 0 & \text{otherwise} \end{cases} \qquad (3.53)$$

which is then normalised by the sum of the template coefficients. Because the transform pair of a Gaussian is also a Gaussian, the sidelobes of the transform do not oscillate as they do for a sinc function and instead just reduce gradually. This implies that the deleterious effects of direct averaging are decreased, though the penalty paid is complexity since the filter now requires floating point operations rather than integer arithmetic. Because averaging gives low-pass filtering, differencing gives a high-pass operation (akin with differentiation in Equation 3.35).

$$\text{difference}(\mathbf{p}[m]) = \begin{cases} \mathbf{p}[m] - \mathbf{p}[m-1] & m = 1, 2 \ldots N - 1 \\ 0 & \text{otherwise} \end{cases} \qquad (3.54)$$

where the result is $N - 1$ samples in duration by virtue of the differencing action (one cannot difference non-existent points). This is a rather rudimentary form of high-pass filtering. Our aim here is not to study signal processing and confine its study to frequency domain properties. The design and analysis of digital filters requires a much deeper analysis and many texts are available [e.g. Ifeachor & Jervis, 2002]. Arguably, the effects of filtering can better be seen with images in the next chapter (because we can see the resulting images, whereas in 1-D we have to play the resulting sounds and because this is a book, playing sound is not available yet in the formats in which this book is published). We shall cover image filtering in Section 4.4.4 to clarify visually the differences between conventional and Gaussian averaging.

3.6 THE FAST FOURIER TRANSFORM

3.6.1 The butterfly operation and basic components of the FFT

3.6.1.1 FFT basis

The DFT would be of little use if there was not an efficient way to compute it. In fact, that is the revolution because of Cooley and Tukey. It is a classic example of computer science: one can achieve efficient computation by considering the structure of an algorithm. This is the *FFT* [Cooley & Tukey, 1965]. Even if it is classic, it is a great pity that an algorithm that brought the power of Fourier analysis to the practical analysis of signals rarely features in Computer Science curricula (OK, it also brought Alexa, but we shan't dwell on that). We could be tempted to digress here, but we shall not. It's a fantastic algorithm and an example of the science in computer science, whereas many just concentrate in the computer bit. Read on!

To approach the algorithm, we first consider the DFT of a set of N points

$$Fp_u = \frac{1}{N} \sum_{x=0}^{N-1} p_x e^{-j\frac{2\pi}{N}ux} \qquad u = 0, 1 \dots N-1$$

The computation is large because the equation involves multiplication by N frequency components for the N points. The frequency components do however have the same structure, and all that changes is their frequency. The key part is then the cosines and sine waves, embodied within the term $e^{-j\frac{2\pi}{N}xu}$. This is where we meet the famous *butterfly* operation shown in Figure 3.24, which looks symbolically like a butterfly (or is supposed to – we regret Gaugin is not in our gene pool) so we popped a real butterfly in. This combines two signals, using an exponential weighting function.

To understand how this relates to the DFT we shall work out the equations that describe all the signals. To do this, we will at times need the relation

$$e^{-j\frac{2\pi}{N}(u+N/2)} = e^{-j\frac{2\pi}{N}(u)} e^{-j\frac{2\pi}{N}(N/2)} = e^{-j\frac{2\pi}{N}u} e^{-j\pi} = -e^{-j\frac{2\pi}{N}u} \qquad (3.55)$$

Bookmark this, as we shall use it. Analysing the process in Figure 3.24, for input points p_0 and p_1, and for $u = 0$ and $N = 2$ the upper of the two-point output is

$$A = p_0 + e^{-j\frac{2\pi}{2}0}p_1$$

because $e^0 = 1$
$$= p_0 + p_1$$

For the lower output $\quad B = p_0 - e^{-j\frac{2\pi}{2}0}p_1$

By our bookmark, $\quad = p_0 + e^{-j\frac{2\pi}{2}\left(0+\frac{2}{2}\right)}p_1$
Equation 3.55

Giving $\qquad\qquad B = p_0 + e^{-j\frac{2\pi}{2}}p_1$

These are the outputs of a two-point DFT since by rephrasing the DFT in Equation 3.6 for two points $N = 2$ (without the scaling factor)

$$Fp_u = \sum_{x=0}^{1} p_x e^{-j\frac{2\pi}{N}ux} \quad u = 0, 1$$

which expands to give the same functions as those of the butterfly

$$Fp_0 = p_0 + p_1 = A \qquad Fp_1 = p_0 + p_1 e^{-j\frac{2\pi}{2}} = B$$

The butterfly operation performs a two-point DFT and there is no computational advantage yet, because we have the same number and

FIGURE 3.24   Butterfly operation for the fast Fourier transform

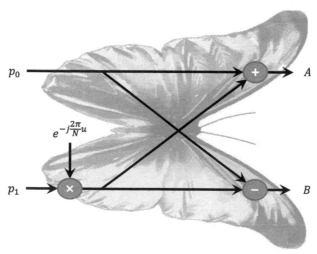

p_0 $\longrightarrow$ $+$ A

$e^{-j\frac{2\pi}{N}u}$

p_1 $\times$ $-$ B

type of computations to perform (both require two additions and one multiplication). The advantage is that one of them can be replicated and one can only be extended: naturally it is the butterfly that can be replicated (as indeed butterflies do). We replicate this by combining two 2-point spectra into a four-point spectrum, in Figure 3.25 forming a kaleidoscope of butterflies (yes, 'kaleidoscope' is one of those multiple words like a charm of goldfinches). Now the first part of this is two 2-point butterflies for which for $u = 0$ and $N = 2$, as before. These butterflies are interleaved, and so the second stage is a four-point transform where $u = 0, 1$ and $N = 4$. Note also the rearrangement of the data that is being transformed, which allows the computation to proceed in this manner. We shall work out how to do this in the next section, for now just go with the flow. This network actually delivers a four-point FFT, but the computation is less than for the DFT. (The papers we write are not diaries, so history does not record precisely how Cooley and Tukey stumbled on the butterfly. It is most possibly by the symmetry, Equation 3.34, and the binary progression of the exponential factors, 'the periodicity of the sine-cosine functions' [Cooley et al., 1967]. The original paper states 'how special advantage can be obtained in the use of a binary computer with $N = 2^m$ ' which is another part of the story.)

Analysing the network in Figure 3.25, we have

$$A = p_0 + e^{-j\frac{2\pi}{2}0}p_2 \text{ and } C = p_1 + e^{-j\frac{2\pi}{2}0}p_3$$

So the top output $\quad E = A + \left(e^{-j\frac{2\pi}{4}0}C\right) = (p_0 + p_2) + (p_1 + p_3)$

and for the next output down $\quad F = B + e^{-j\frac{2\pi}{4}1}D$

because $B = p_0 - e^{-j\frac{2\pi}{2}0}p_2$ and $D = p_1 - e^{-j\frac{2\pi}{2}0}p_3$

then $\quad F = B + e^{-j\frac{2\pi}{4}1}D$

by substitution for B and D $\quad = \left(p_0 - e^{-j\frac{2\pi}{2}0}p_2\right) + e^{-j\frac{2\pi}{4}1}\left(p_1 - e^{-j\frac{2\pi}{2}0}p_3\right)$

by Equation 3.55 $\quad = \left(p_0 + e^{-j\frac{2\pi}{2}\left(0+\frac{2}{2}\right)}p_2\right)$

$$+ e^{-j\frac{2\pi}{4}1}\left(p_1 + e^{-j\frac{2\pi}{2}\left(0+\frac{2}{2}\right)}p_3\right)$$

expanding gives
$$= p_0 + e^{-j\frac{2\pi}{4}1} p_1 + e^{-j\frac{2\pi}{2}\left(0+\frac{2}{2}\right)} p_2 + e^{-j\frac{2\pi}{4}1} e^{-j\frac{2\pi}{4}(2)} p_3$$

and finally (phew!) $F = p_0 + e^{-j\frac{2\pi}{4}1} p_1 + e^{-j\frac{2\pi}{4}2} p_2 + e^{-j\frac{2\pi}{4}3} p_3$

Let us do it in G (that's a line from the Monty Python musical Spamalot, in the song that never ends). Because $C = p_1 + e^{-j\frac{2\pi}{2}0} p_3$

$$G = A - e^{-j\frac{2\pi}{4}0} C = p_0 + p_2 - e^{-j\frac{2\pi}{4}0}\left(p_1 + e^{-j\frac{2\pi}{2}0} p_3\right)$$

Because $e^{-j\frac{2\pi}{2}0} = e^{-j\frac{2\pi}{4}4}$
$$= p_0 + p_2 + e^{-j\frac{2\pi}{4}\left(0+\frac{4}{2}\right)}\left(p_1 + e^{-j\frac{2\pi}{4}4} p_3\right)$$

$$= p_0 + e^{-j\frac{2\pi}{4}2} p_1 + p_2 + e^{-j\frac{2\pi}{4}2} e^{-j\frac{2\pi}{4}4} p_3$$

Because $1 = e^{-j\frac{2\pi}{4}4}$ $G = p_0 + e^{-j\frac{2\pi}{4}2} p_1 + e^{-j\frac{2\pi}{4}4} p_2 + e^{-j\frac{2\pi}{4}6} p_3$

We could do this for H and we have, but we won't put it here as that would be sadism. So let us instead relate this back to the DFT. By Equation 3.6, a four-point DFT is

$$Fp_u = \sum_{x=0}^{3} p_x e^{-j\frac{2\pi}{4}ux}$$

so

$$Fp_0 = p_0 + p_1 + p_2 + p_3 = E$$

FIGURE 3.25 Combining two 2-point spectra

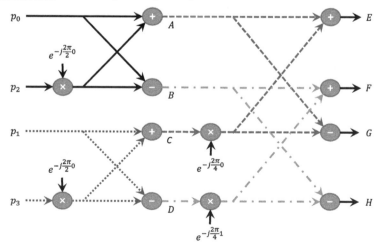

$$Fp_1 = p_0 + p_1 e^{-j\frac{2\pi}{4}1} + p_2 e^{-j\frac{2\pi}{4}2} + p_3 e^{-j\frac{2\pi}{4}3} = F$$

$$Fp_2 = p_0 + p_1 e^{-j\frac{2\pi}{4}2} + p_2 e^{-j\frac{2\pi}{4}4} + p_3 e^{-j\frac{2\pi}{4}6} = G$$

$$Fp_3 = p_0 + p_1 e^{-j\frac{2\pi}{4}3} + p_2 e^{-j\frac{2\pi}{4}6} + p_3 e^{-j\frac{2\pi}{4}9} = H$$

These are derived from the original DFT and are the same as E, F, G and H, respectively.

So far, the systems here have assumed $N = 2^M$ where M is an integer (a.k.a. N is an integer power of 2). This is not always the case, but it is usually easy to zero-pad the data by adding points that are zero until the length of the vector is 2^M.

3.6.1.2 FFT computation and speed

The system in Figure 3.25 leads to an FT of the four input points, when they have been arranged in a special way. So what is all the fuss? Well, we can work out the numbers of additions and summations. In the four-point DFT there are nine multiplications and four additions. In the four-point FFT in Figure 3.25 there are four multiplications and eight additions. In an eight-point DFT there are 49 multiplications and 7 additions. If we were to double the size of Figure 3.25 the eight-point FFT would have 8 multiplications and 16 additions to which are added an extra four multiplications and eight additions giving in total 12 multiplications and 24 additions. These are summarised in Table 3.1, which shows the continuing reduction in FFT computation compared with that for the DFT, for larger values of N. In general, in terms of multiplications, in the N-point DFT there are $(N-1)^2$ multiplications and N^2 if the zero-order terms are considered. For later use, we shall state

$$\text{the computational cost of the DFT} = O\left(N^2\right) \qquad (3.56)$$

Let us now see what the computational cost/*complexity* of the FFT actually is, in general. The time taken changes with the size of the data used, N. If we consider the butterfly operation, or the more generalized version in Figure 3.25, then its calculation for four points needs two complex multiplications and four additions (subtractions cost the

TABLE 3.1 Computational cost of different DFT and FFT sizes.

2-point DFT		2-point FFT		4-point DFT		4-point FFT		8-point DFT		8-point FFT	
+	×	+	×	+	×	+	×	+	×	+	×
2	1	2	1	4	9	8	4	7	49	24	12

same in two's complement) and the two stages of two-point trans-forms. As in [Mallat, 2008] the cost of a N point transform, $C(N)$, cal-culated from two $N/2$ transforms (as in Figure 3.25) is

$$C(N) = 2C(N/2) + kN \qquad (3.57)$$

where the term kN is the extra arithmetic cost of that stage. Now if we consider the FFT to be constructed of a series of levels, each of which has $N = 2^l$ points, then $l = \log_2(N)$. In Figure 3.25 we have levels $l = 0$, 1 (the 2-point FFT) and $l = 2$ (the 4-point FFT). The proportionate cost of each level is $CL(l)$

$$CL(l) = \frac{C(N)}{N} \qquad (3.58)$$

So by substitution in Equation 3.57

$$CL(l) = CL(l-1) + k$$

(The proportionate cost at level l includes two stages of the preceding level $l-1$, each of which has half the arithmetic.) The FT of a single point is itself, so $C(0) = CL(0) = 0$ and the cost depends on the number of levels $CL(l) = k \times l$, which by Equation 3.58 gives

$$C(N) = kN \log_2(N) \qquad (3.59)$$

We shall consider k again later, and for now we shall state

$$\text{the computational cost of the FFT} = O\left(N \log_2 N\right) \qquad (3.60)$$

(which is in the original description [Cooley & Tukey, 1965] but stated as $O(N \log N)$ – not base 2: it's obvious!). Given that N is usually large and $\log_2 N$ is much smaller, this is much (and much) faster than the DFT by the factor

$$\frac{\text{the computational cost of the FFT}}{\text{the computational cost of the FT}} = \frac{O\left(N \log_2 N\right)}{O\left(N^2\right)} = O\left(\frac{\log_2 N}{N}\right) \qquad (3.61)$$

We shall return to computational matters later, since they motivated the development of the FFT.

There is current debate about speed of additions relative to that of multiplications. This largely depends on the computational archi-tecture and the numbers themselves. In integer arithmetic, addition and subtraction are the same speed because of the use of comple-ment arithmetic (for digital arithmetic, no place like home: try [Nixon, 2015]). Multiplication of integers can take longer because of a shift and then add structure. For floating point numbers, addition requires nor-malisation of the mantissas until the exponents match prior to addi-tion, whereas floating point multiplication requires multiplication of the mantissas and addition of the exponents. Thus a floating-point

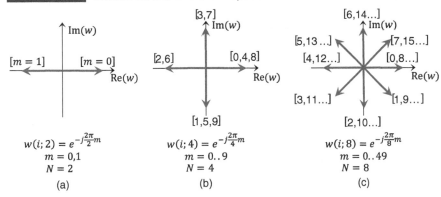

FIGURE 3.26 Twiddle factors and their repetitions

$$w(i;2) = e^{-j\frac{2\pi}{2}m}$$
$$m = 0,1$$
$$N = 2$$
(a)

$$w(i;4) = e^{-j\frac{2\pi}{4}m}$$
$$m = 0..9$$
$$N = 4$$
(b)

$$w(i;8) = e^{-j\frac{2\pi}{8}m}$$
$$m = 0..49$$
$$N = 8$$
(c)

multiplication can be faster than addition. But there are also implications by the cache (local storage), the use of look-up tables (particularly in division) and the use of dedicated hardware. The consensus remains that the computational cost of the FFT is O $(N \log_2 N)$, but if one is to go into the detail – as we do later – it is much more complicated than that. As we shall find, it is not so difficult to build the full FFT and standard implementations are available. So if speed is of particular concern, build a selection of algorithms and time them, then choose the fastest one. We shall return to computational matters in Section 3.6.4 and consider optimised FFT algorithms in Section 3.6.5. Before that we have to finish off building the FFT, first by making it larger than for four points.

3.6.1.3 Extending the FFT

The factors that include the cosine and sine waves, $e^{-j\frac{2\pi}{N}m}$ (for $m = ux$), are actually called *twiddle factors*. An alternative notation for a twiddle factor w is much simpler

$$w(m;N) = e^{-j\frac{2\pi}{N}m} \tag{3.62}$$

though using the notation $w(m;N)$ omits clear relation to frequency. We shall use both in the analysis that follows. For the two-point transform, $N = 2$, there are two orientations and both are along the real axis, Figure 3.26(a). For the four-point transform, we add two imaginary orientations as in Figure 3.26(b), which can also repeat, where $w(0;4) = w(4;4)$ et seq. For the eight-point transform, there are eight orientations at angles of 45° as in Figure 3.26(c). The repetition is a consequence of the succession of twiddle factors in successive stages of the transform. We shall use this in the mathematical analysis of the operations here.

FIGURE 3.27 Four-point FFT

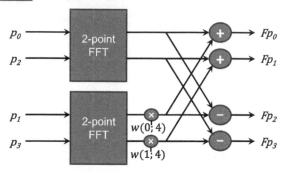

Using the compact notation for the twiddle factors in Equation 3.62, we can derive a simpler version of the four-point FFT as shown in Figure 3.27.

The implementation of the four-point FFT of Figure 3.27 is given in Code 3.3. This treats the inputs as scalars and delivering an output vector of Fourier coefficients (but note that the indices of the output vector are from 1 to 4, not 0 to 3 as they are in the maths).

Code 3.3 The 4-point FFT

```
function Fp4 = FourPointFFT(p0,p2,p1,p3)
%The 4-point FFT from Fig 3.27

%Fp(n) = upper butterfly wing
%           +(lower butterfly wing * twiddle factor)
Fp4(1) = p0+exp(-j*2*pi*0/2)*p2
           + (p1+exp(-1j*2*pi*0/2)*p3)*exp(-j*2*pi*0/4);
Fp4(2) = p0-exp(-j*2*pi*0/2)*p2
           + (p1-exp(-1j*2*pi*0/2)*p3)*exp(-j*2*pi*1/4);
Fp4(3) = p0+exp(-j*2*pi*0/2)*p2
           - (p1+exp(-1j*2*pi*0/2)*p3)*exp(-j*2*pi*0/4);
Fp4(4) = p0-exp(-j*2*pi*0/2)*p2
           - (p1-exp(-1j*2*pi*0/2)*p3)*exp(-j*2*pi*1/4);
end
```

We shall put these together to give an eight-point FFT as in Figure 3.28.

The eight-point FFT is implemented in Code 3.4, again noting that Matlab's indices go from 1 to 8, not 0 to 7 and the FourFp_upper and FourFp_lower are the four-point FFTs in Figure 3.28. This uses the four-point FFT function in Code 3.3 and so the data is rearranged to be in the order required for that operation.

FIGURE 3.28 Eight-point FFT

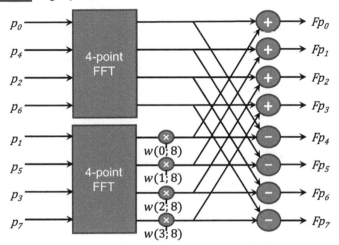

Code 3.4 The 8-point FFT

```
function Fp = EightPointFFT(p)

FourFp_upper=FourPointFFT(p(1),p(5),p(3),p(7));
%transform rearranged data
FourFp_lower=FourPointFFT(p(2),p(6),p(4),p(8));
%transform rearranged data
for m=1:4
  Fp(m) = FourFp_upper(m)
              +FourFp_lower(m)*exp(-1j*2*pi*(m-1)/8);
  Fp(m+4) = FourFp_upper(m)
              -FourFp_lower(m)*exp(-1j*2*pi*(m-1)/8);
end
```

An alternative way to check the FFT algorithms is to use the inverse DFT/FFT to check we have returned to the original place. This does not however check we went to the right place, just that we came back from it to the same place. Another way is to transform a sampled pulse and check values are the same as those calculated from the sinc function in Equation 3.18.

So we shall apply the eight-point FFT from Code 3.4 in comparison with Matlab's own fft, and with the DFT, as shown in Code 3.5. There are eight data points stored in a vector p and from these we form the DFT – which is multiplied by 8 since there is no scaling/normalisation by $1/N$ in Matlab's fft. The Fourier data are presented as truncated absolute values, for compactness only (which is the same reason as using the notation [...] for output values since this is not within Matlab's grammar either). The eight points resulting from the FFT appear to be the same as those from the DFT and there are numerical errors $O(10^{-12})$ which cannot be seen. The same result ensues for the eight-point FFT of Code 3.4 and it is not shown here since it would be pure repetition. The difference between the eight-point FFT and the DFT is $O(10^{-12})$ since they perform the same computation in different ways and the difference between fft and the eight-point FFT is zero. For comparison of the results, the outputs of the DFT are taken as the 'true' values. We now need to work out how to order the data prior to transformation.

Code 3.5 Analysing the FTs

Data

p=[10 20 30 40 50 60 70 80];

DFT×8 (all numbers rounded for display)

abs(DFT(p)*8) = [360 105 56.6 43.3 40.0 43.3 56.6 105]

Matlab's fft

abs(fft(p)) = [360 105 56.6 43.3 40.0 43.3 56.6 105]

difference between DFT and 8-point FFT

abs(DFT(p)*8)-abs(EightPointFFT(p))
            = 1.0e-12 * [0 0 0 0.02 0 0.14 0.16 −0.21]

difference between fft and 8-point FFT

abs(fft(p))-abs(EightPointFFT(p))
            = 0 0 0 0 0 0 0 0

3.6.2　Decimation in time

The data needs to be arranged in a special way for the butterfly operations, as in Figure 3.28. We shall find this can easily be achieved. Considering the DFT

$$Fp_u = \frac{1}{N} \sum_{x=0}^{N-1} p_x e^{-j\frac{2\pi}{N}xu}$$

By splitting the summation into two halves where each contains $N/2$ points, we have

$$Fp_u = \frac{1}{N}\sum_{x=0}^{\frac{N}{2}-1} p_{2x}e^{-j\frac{2\pi}{N}2xu} + \frac{1}{N}\sum_{x=0}^{\frac{N}{2}-1} p_{2x+1}e^{-j\frac{2\pi}{N}(2x+1)u}$$

$$= \frac{1}{N}\sum_{x=0}^{\frac{N}{2}-1} p_x^e e^{-j\frac{2\pi}{N}\frac{2xu}{2}} + e^{-j\frac{2\pi}{N}u}\frac{1}{N}\sum_{x=0}^{\frac{N}{2}-1} p_x^o e^{-j\frac{2\pi}{N}\frac{xu}{2}}$$

(3.63)

where p_x^e and p_x^o are the even-numbered and the odd-numbered parts of the data, respectively. By inspection, Equation 3.63 is the addition of two DFTs of the $N/2$ even-numbered and odd-numbered points

$$\mathcal{F}(\mathbf{p}, N) = \frac{1}{2}\mathcal{F}(\mathbf{p}^e; N/2) + \frac{1}{2}e^{-j\frac{2\pi}{N}u}\mathcal{F}(\mathbf{p}^o; N/2) \qquad (3.64)$$

By Equation 3.63 we can recursively split the data into odd-numbered and even-numbered parts. We could do this by maths based on Equation 3.64, but the summations would become increasingly complex (there is a very neat analytical description in [Mallat, 2008] of the FFT based on a form of Equation 3.64, though its implementation is less transparent). We shall use a diagram instead, Figure 3.29. Let us take eight data points $p_0..p_7$ and we shall first split them into even (e) and odd (o) parts which each contain half the points as in Equation 3.63. This is level 1, which contains four odd and four even data points. Now we can repeat the operation, to give Level 2, which has four groups each containing two data points. The path is given by the split to achieve this grouping so the label 'ee' in Level 2 refers to the even part of the even group in Level 1. This can be repeated to give

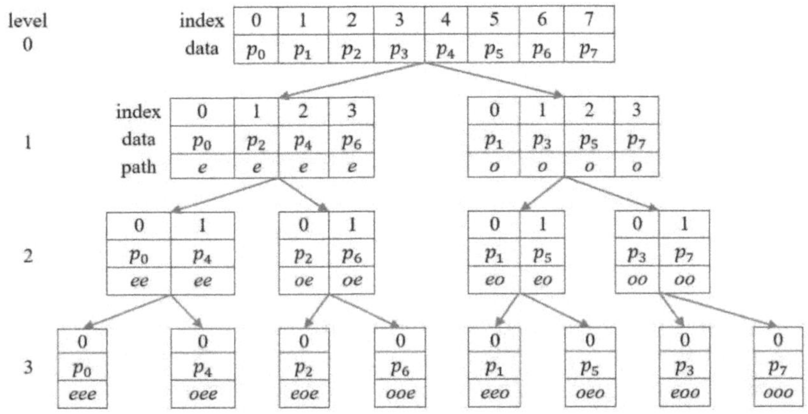

FIGURE 3.29 Decimating data for the fast Fourier transform

the 8 groups of a single point in Level 3, and the paths refer to the three splits taken to achieve this grouping. The points in Level 3 are in the same order as in Figure 3.28.

This process is called decimation in time and is a recursive process. If one is to replace 'e' with '0' and 'o' with '1' then the path can be decoded as the binary index of each data point: the path to p_0 is 'eee', which is 000 in binary, and the path to p_5 is 'oeo', which is 101. So we now have an efficient way of splitting up the data and this can be used to reconstruct the original sequence of data points.

The implementation of this is given in Code 3.6. This splits the data successively into half even and half odd, dropping out in the last level. We shall not optimise this code since all that is needed in this educational excursion is to show that it can easily be done, as it indeed is.

Code 3.6 Decimating data in time

```
function Np = decimate_time(p)

%This prepares the data for the N-point FFT

cols=length(p);
Fp(1:cols)=0;
N=log2(cols); %assume we have 2^N data points
Np=p; %storage has not been optimised;
     %we will need intermediate storage
%iterate until log2(cols) levels have been performed
split=cols/2; %start at half of the data
while split>1 %1 is the bottom level
  for j=1:2*split:cols
    for i=1:split %spit into even and odd
      even(i)=Np(j+(i-1)*2);
      odd(i)=Np(j+(i-1)*2+1);
    end %now put them in the original array
    Np(j:j+split-1)=even(1:split);
    Np(j+split:j+2*split-1)=odd(1:split);
  end
split=split/2; %and go and do the next level (down)
end
```

We shall use Code 3.6 to decimate data in Code 3.7, which shows decimation for $N = 2, 4$ and 8. The points for $N = 8$ are again in the same order as in Figure 3.28 and as in Level 3 in Figure 3.29.

Code 3.7 Decimating data

```
p = [0 10 20 30]
decimate_time(p) = [0 20 10 30]
p = [0 10 20 30 40 50 60 70]
decimate_time(p) = [0 40 20 60 10 50 30 70]
p = [0 10 20 30 40 50 60 70 80 90 100 110 120 130 140 150]
decimate_time(p)
    = [0 80 40 120 20 100 60 140 10 90 50 130 30 110 70 150]
```

The most common form of decimation in time is the bit reversal algorithm. In outline, given a conventional binary number written from Most Significant Bit (MSB) to Least Significant Bit (LSB) as MSB...LSB then bit reversal reorders the bits from LSB...MSB. The bits are reversed in position, not inverted in magnitude. For decimal numbers 0...7 the process is shown in Table 3.2.

The resulting bit reversed numbers in decimal are the same as the order in Code 3.7, with $N = 8$ in Figure 3.28, and as in Level 3 in Figure 3.29. An example bit reversal for a point within a 16-point vector (and thus with a 4-bit binary index) is shown in Figure 3.30. As shown in Code 3.7, after bit reversal the new position of p_{13} is p'_{11}.

The process appears routine, though as ever there is interest in fast algorithms: reviews include software [Karp, 1996] and hardware

TABLE 3.2 Bit reversal.

	Decimal	0	1	2	3	4	5	6	7
	Binary	000	001	010	011	100	101	110	111
Binary bit reversed		000	100	010	110	001	101	011	111
Binary bit reversed in decimal		0	4	2	6	1	5	3	7

FIGURE 3.30 Example bit reversal in 16-Point FFT data (for p_{13})

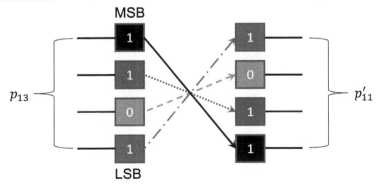

[Garrido et al., 2011]. The interest continues with work in parallel implementations [Cheng & Yu, 2015].

3.6.3   Radix 2 FFT

We can now extend Code 3.4 to give the full FFT in Code 3.8. This will use the function `decimate_time(p)` to prepare the data before transformation. The code implements the *radix 2 FFT* because that is its basis. Essentially the code builds from the two-point butterflies in the first level, to the four-point transform in Figure 3.27 and then to the eight-point one in Figure 3.28, et seq. The implementation of the radix 2 FFT is compact, largely since it is a succession of butterfly operations at each level L. The outputs of the butterfly operations are temporarily stored in vectors `upp` and `low` which then form the transform basis for the next level up. The only nasty bit of the code is the calculation of the pointers, but these largely use integer arithmetic. Essentially, the code is only slightly more complex than that for the DFT, Code 3.1, since half the code is copying the data from its temporary storage. The code could be optimised but it is aimed to be clear and to parallel the presentation here of the development of the FFT (this is the Educational FFT!). For that reason it omits zero-padding (assuming that the input vector p contains 2^N decimated data points), decimation and windowing.

Code 3.8 Radix 2 FFT

```
function Fp = radix2FFT(p) %This is the N-point Radix 2 FFT

cols=length(p);
Fp(1:cols)=0;
L=1; %assume we have 2^N decimated data points, N>0.
%Start at the bottom
Fp=p; %storage has not been optimised.
%we will need intermediate storage
while L<cols
%iterate until log2(cols)-1 levels have been performed
   for j=1:2*L:cols %do all the points in L/2 batches
    for i=1:L %now do L butterflies
       upp(((j+1)/2)+i-1)= Fp(j+i-1)+
            Fp(j+L+i-1)*exp(-1j*2*pi*(i-1)/(L*2));
       low(((j+1)/2)+i-1)= Fp(j+i-1)-
            Fp(j+L+i-1)*exp(-1j*2*pi*(i-1)/(L*2));
    end
   end
```

```
   for j=1:2*L:cols %copy the components across,
                    %to the right places
      for i=1:L
        Fp(j+i-1)=upp(((j+1)/2)+i-1);
        Fp(j+L+i-1)=low(((j+1)/2)+i-1);
      end
   end
L=L*2; %and go and do the next level (up)
end
```

FIGURE 3.31 Radix 2 FFT

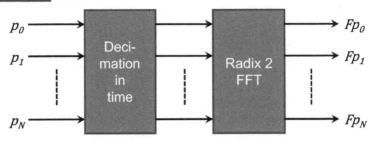

The full architecture of the radix 2 FFT is shown in Figure 3.31 which is invoked by `transformed = radix2FFT(decimate_time(p))`. For data that is not of length 2^N then the command `transformed = radix2FFT(decimate_time(zero_pad(p)))` should be used instead.

This code was checked in the same manner as the eight-point FFT: given a much larger input vector containing 16384 points from the Hard Day's Night chord. Code 3.9 checks that the outputs of the DFT, FFT and radix 2 FFT differ only by numerical error. This was achieved by averaging the values of the difference between the absolute values of each transform. The smallest average difference is between Matlab's FFT and the radix 2 FFT which is $O(10^{-20})$ whereas the differences between the DFT and the two FFTs are $O(10^{-18})$. All figures represent numerical error, though one is around 500 times smaller. This is to be expected since the FFT is a rearrangement of the DFT, for speed (and it certainly achieves that, as we shall find in the next section).

Code 3.9 FFT operation and differences

```
hard=audioread('hard_chord_long.wav');
p=hard(1:2^14); %let's take 16384 points
N=length(p); %to normalise the differences
d=DFT(p); %calculate DFT (takes 1 minute, machine dependent)
f=fft(p)/N; %FFT with scaling, for similarity (takes <<1 sec)
```

```
r=radix2FFT(decimate_time(p))/N;
%radix 2 FFT with scaling, for similarity
sumdf=0; sumdr=0; sumfr=0; %initialise summations
for u=1:N
    sumdf=sumdf+(abs(d(u))-abs(f(u))); %difference DFT-FFT
    sumdr=sumdr+(abs(d(u))-abs(r(u)));
                %difference DFT-radix 2 FFT
    sumfr=sumfr+(abs(f(u))-abs(r(u)));
                %difference FFT-radix 2 FFT
end
sumdf/N %average difference DFT-FFT
ans = 7.7426e-18
sumdr/N %average difference DFT-radix 2 FFT
ans = 7.7585e-18
sumfr/N %average difference FFT-radix 2 FFT
ans = 1.5915e-20
```

There are other implementations that are publicly available. For example there is a selection of algorithms available in many different languages https://rosettacode.org/wiki/Fast_Fourier_transform but it does not contain a Matlab version since 'Matlab/Octave have a built-in FFT function'. Clearly there is no educational agenda there, though there is a welcome pragmatic one nonetheless. There is a pseudocode implementation in Wikipedia which descends the butterfly stack, recursively, and has a similar complexity to the code in Code 3.8. It is more elegant, via recursion, but uses more memory for the same reason. Code 3.8 does not follow this, as it follows the logical progression from the 2-point butterfly. There are split radix implementations, to be discussed later since the desire to apply and continuing use of the FFT motivates interest in improving speed by optimisation. Let us first look at computational cost since that was the prime motivation for developing the FFT.

3.6.4 Computational time for FFT compared with DFT

3.6.4.1 Improvement in speed vs DFT

In Section 3.6.1.2 we started to analyse computation of the FFT. Now we have the complete transform we shall compare the time required to calculate the FTs for different values of N, shown in Figure 3.32, with timings obtained using Matlab's tic toc functions. All values of N for the data vectors are integer powers of 2, $N = 2^M$. We shall compare the times taken by the DFT from Code 3.1, Matlab's own fft and the radix 2 FFT from Code 3.8. In Figure 3.32(a) the execution time for the DFT can be seen to increase dramatically as *N* increases and the two FFTs take only a fraction of the time (the two FFTs appear superimposed here). For 4096 points, the DFT took around 4s (which is about the time it took you to read this sentence) whereas the radix 2 FFTs took 0.02 s (the time to read just one letter). There is no need to extend the graph to include the computation for 8192 points: the same shape of graph would ensue. The radix 2 FFT is slower than Matlab's `fft` since Matlab's `fft` is compiled rather than interpreted (line by line). For 4096 points (real points from the sample of a Hard Day's Night, Figure 3.13), on average the FFT took 0.0000596 s, the radix 2 FFT took 0.0183 s and the DFT 4.23 s. So on average the radix 2 FFT here is more than 200 times faster than the DFT. Matlab's `fft` here is more than 300 times faster than the radix 2 FFT and around 70,000 times faster than the DFT. All times were averaged over multiple runs (as the processor was occasionally diverted by looking at the World Cup cricket scores) and the variance is not shown (the maximum variance was .002s which would not show).

Let us summarise where we got to in Section 3.6.1.2 in key point 7.

The fast Fourier transform (FFT) gives the same result as the discrete FT and for *N* points it is $\dfrac{N}{log_2 N}$ times faster (and when *N* is large that means a lot faster).

To investigate whether the computational cost of the radix 2 FFT is actually O $(N log_2 N)$, as predicted in Equation 3.60, we have also plotted the ratio of the cost of the DFT (N^2) (Equation 3.56) divided by the

FIGURE 3.32 Performance of DFT and FFT algorithms: (a) execution time; (b) actual and predicted fractional performance

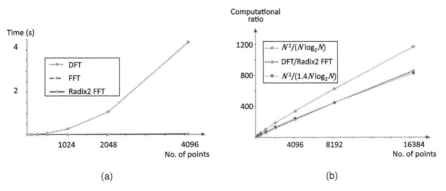

(a)

(b)

cost of the radix 2 FFT ($Nlog_2N$), in Figure 3.32(b). These are both programmed in Matlab code, and so the comparison is a fair one (Matlab's fft is intrinsically faster). Then we have superimposed the ratio of the actual cost of the DFT divided by the actual cost of the radix 2 FFT. If the cost of the radix 2 FFT was actually $Nlog_2N$ then the traces would be the same and they are not. The ratio DFT/ Radix2 FFT is less than $Nlog_2N$ – i.e. the cost of the DFT is less (yes – if the zero-order terms are discounted, but otherwise unlikely) or the cost of the FFT is more than O $\left(Nlog_2N\right)$. As it is more likely the cost of the FFT is greater since the estimated cost is largely based on the number of multiplications, Figure 3.32(b) also includes the curve for O $\left(1.4Nlog_2N\right)$, which over this interval – and for this data and this implementation – matches better the curve for what is actually achieved in Matlab (and substantiates Equation 3.59). More importantly the curves have a similar shape: the computational cost of the FFT is indeed O $\left(Nlog_2N\right)$ and the O matters! There is a much deeper and more sophisticated analysis of FFT algorithms (between 1968 and 2006) that is publicly available http://www.fftw.org/benchfft/ measuring performance and accuracy on different architectures.

This is excellent computer science: for slightly more complex code we can derive an enormous speed advantage whilst still getting the same result. The FFT can be used within Short-Time Fourier, taking advantage of its enormous speed advantage. The development of the FFT means that we can use discrete Fourier analysis in applications, without it we could not.

3.6.4.2 Speeding convolution via the convolution theorem

So is the FFT always faster in applications? This is not necessarily true, since we need to call the FFT function and iterate through its code. We shall consider circular convolution (avoiding the zero-padding of linear convolution) implemented either directly or using the convolution theorem. The question is then 'for what size of data is a direct implementation of convolution faster than one which uses the Fourier transform?'. One text [Smith, 2011] found that by using a MATLAB script 'FT convolution was found to be faster than direct convolution starting at length $N = 2^6 = 64$' and this is repeated elsewhere on the Web. Another text [Strum88] suggests $N = 2^7 = 128$ by (theoretical) comparison of the number of real multiplications used. Let's take a look.

We shall presuppose that one of the transforms is precomputed (say the system response) since it is fixed. Thus using Fourier the process is inverse_radix2_FFT(radix2_FFT(p)*q) (where the two FFT's are based on Code 3.8 and q is the precomputed transform); for direct convolution we use Code 3.2. Both are thus direct Matlab code and Matlab's fft is not used as it is pre-compiled. For N data points, the computational cost of the 1-D FFT based implementation is of the order of $N \log_2 N$ and two transforms need to be multiplied together. The total cost of the Fourier implementation of circular convolution is then

$$\text{Cost}_{\text{FFT}} = O\left(2N \log_2 N + N\right)$$

The cost of the direct implementation of circular convolution for N points is N^2 multiplications for each point. These multiplications are real which requires less time than for a complex multiplication within the transform-based approach. The (comparative) cost of the direct implementation of convolution is of the order of

$$\text{Cost}_{\text{direct}} = O\left(N^2\right) \tag{3.65}$$

There are higher considerations of complexity than this analysis has included, including decimation in time. One calculation for M non-zero data points calculates the computational cost as $O\left(3M \log_2 M + 11M\right)$ thus giving faster implementation of circular convolution by the FFT for $M \geq 32$ [Mallat, 2008]. To resolve all this, let us look at what the code does. Via the Matlab implementations, shown in Figure 3.33(a), the speed advantage of the direct implementation is only apparent for small sets of data; for more reasonably sized sets of data > 256 the FFT implementation is clearly much faster. A check on the theoretical predictions based on Equation 3.65 and using the cost of multiplications calculated by [Strum & Krik, 1988] is shown in Figure 3.33(b) as the ratios of the cost of the direct implementation to the FFT based one. This suggests that the advantage found in practice can be confirmed by consideration of the implementation. Again, the Fourier costs have a similar trajectory but neither plot matches exactly the

FIGURE 3.33 Performance DFT and FFT implementations of circular convolution: (a) execution time; (b) cost of direct/ FFT for theory and practice

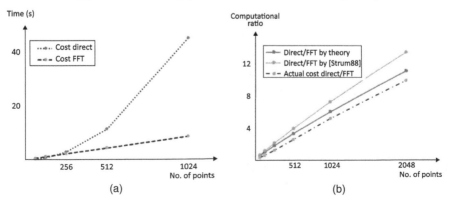

ratio that is found with the Matlab implementations. There are, naturally, further considerations in the implementation (the cost of which has been omitted here) and in the use of transform calculus, the most important being the use of windowing. This implies that implementing convolution using transform calculation should be used when larger sizes of data are involved and the number of points exceeds 128, and when speed is critical. In a sense it is academic, since it is only for small values of N that the direct implementation is fastest, but there might be batch-processing situations where this is important. Certainly, the developments in Deep Learning have employed convolution within their basis, motivating investigations of efficient implementation and thus the FFT. There are approaches [Gu et al., 2018; Lavin & Gray, 2016] (in the focus of this Chapter, for speech recognition) based on Winograd's *minimal filtering algorithm* which exploit computational structure to improve speed – we'll hear of Winograd again when considering even faster FFT algorithms in Section 3.6.6.

3.6.5 Optimising the FFT

So it is certainly fast, but how can the FFT be made even faster? Matlab code is interpreted (line by line) whereas Matlab's `fft` is compiled and that makes an enormous difference, especially with regular data structures as used in the FFT. We could never optimise Matlab code and catch up with Matlab's own `fft`. We could exploit parallelism since the FFT is clearly a parallel algorithm [e.g. Gupta & Kumar, 1993]. We could implement the FFT purely in hardware, and the architecture is already given (and there is even relatively recent interest [e.g. Ayinala & Parhi, 2013; Ma et al., 2015; Salehi et al., 2013]). There are many minor modifications that can be made to the code (notwithstanding Matlab's own advice to avoid the variable length vectors – here, `upp` and `low` – for reasons of speed). By Figure 3.26 the twiddle factors $w(0;2) = 1$ and $w(1;4) = -j$ and this happens for different values of m and N (in $w(m;N)$). It is then possible to write the four-point transform via Figure 3.26 as

$$Fp = \begin{bmatrix} 1 & 1 & 1 & 1 \\ 1 & w(1;4) & -w(0;2) & -w(0;2)\,w(1;4) \\ 1 & w(0;4) & w(0;2) & -w(0;2)\,w(0;4) \\ 1 & -w(1;4) & -w(0;2) & w(0;2)\,w(1;4) \end{bmatrix} p$$

$$= \begin{bmatrix} 1 & 1 & 1 & 1 \\ 1 & -j & -1 & j \\ 1 & -1 & 1 & -1 \\ 1 & -j & -1 & j \end{bmatrix} p$$

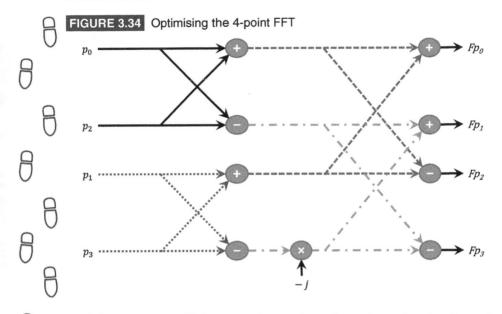

FIGURE 3.34 Optimising the 4-point FFT

and this gives an efficient 4×4 transform kernel previously shown in Equation 3.8. Then we can use the symmetry in the FFT since Fp_1 is the complex conjugate of Fp_{N-1}. These are shown for an optimised four-point FFT in Figure 3.34 and the architecture is clearly much simplified (three multiplications are avoided). We have not used any of these optimisations since it would be pointless in Matlab. We shall use Matlab's `fft` in future since it is faster and delivers the same result.

3.6.6 Even faster FFT algorithms

The FFT implementation here is educational only. In a practical sense it is futile because it is written in Matlab code and Matlab has its own FFT already. But Matlab's own FFT had to come from somewhere, and it must be good or people would use a different one. It is then no surprise that there is continuing interest in even faster FFT algorithms. One of the dichotomies in applied science is that ambitions invariably exceed the means to accomplish them: with faster processing we can handle more data and yet we invariably want to process even more. As such the interest in optimisation continues now, and will always do so in the future.

Essentially, optimising the FFT concerns a trade-off between addition, multiplication and storage. This is a moving target, since computational architectures evolve, and there is the additional extension to multi-thread capability on modern computers. Some of the implementations might use dedicated architectures such as Field Programmable Gate Arrays (FPGAs), Digital Signal Processing (DSP) chips or RISC or graphics processors (GPUs) for more regular data. The selection of strategies include Prime Factor Algorithms [Kolba & Parks, 1977], the Winograd FFT [Winograd, 1978] and split radix approaches [e.g. Johnson & Frigo, 2006]. The water is muddied more by (further complicated by) dedicated processing requirements, such as for convolution. There is no optimum choice: the end justifies the means, and the end is often an application. Engineering is ever about compromise to reach an objective.

One of the fastest publicly available implementations is called FFTW, which derives from the superb name Fastest FT in the West http://www.fftw.org/. This achieves speed by using a set of neat software tricks, which optimises the implementation according to the architecture used, and the data to be analysed [Frigo & Johnson, 2005]. The approach involves planning to describe the arrangement of input data which adapts to the hardware and execution of the processing system, giving a plan to execute the transform. The planner measures the run time of different plans thus allowing selection of the fastest available. It is very neat and beautifully described. (There is an fftw function in Matlab, which cites the address above, though it does not mention 'West'.) There is also an eastern counterpart, FFTE http://www.ffte.jp/, which appears to offer speed advantage for large data sets [Nikolic et al., 2014].

Interest in optimising the speed of the FFT continues, as it is a baseline algorithm and architectures change. We shall not survey the more modern approaches, since any review would be out of date too quickly for this text. Computers change, and so do algorithms. After all, quantum computing is around the corner.

3.7 SUMMARY

Pretty much everything we do now uses computers, so in this chapter we moved from continuous signals to sampled ones. Perhaps surprisingly, the sampling process is more complex than just taking the points that represent a signal. If we take them often enough, sample at a high frequency, then the samples are an adequate representation. When the sampling frequency is too low, aliasing can corrupt the signal and we can get completely the wrong thing. This leads to the Nyquist sampling theorem which states that signals should be sampled at least at twice their maximum frequency. When we have sampled a signal, we can derive a DFT, and its inverse form for reconstruction. As with its continuous version, there are basic signals and properties that allow the analysis of the DFT. Again there are discrete equivalents to processing signals, including correlation, convolution and filtering. The major advantage of the discrete version is that there is a fast version, the FFT, which allows for much faster processing. It turns the transform from one which is tediously slow into one which is blindingly fast, without much extra software. It's neat stuff indeed. The FFT is based on a simple notion and is not difficult to implement and has been shown experimentally to be much faster. We shall need the FFT when we process signals in two dimensions, images, where we can see the results rather than just hear them as in the 1-D signals in this chapter. We have not considered here the newer approaches that can handle irregular data which are known as *graph signal processing* [Ortega et al., 2018]. For example Twitter/ Facebook/ Linkedin could represent data which includes people, their connections and networks, and this can extend to more conventional, though irregular, representations of electrical stimuluses and this has implications on sampling, as mentioned earlier, and on transforms.

Where we are now is the DFT allows transformation and the FFT makes it fast. For images, the FFT allows transformation in a reasonable time, whereas with the DFT we'd just have to wait too long (or buy fancy computers). On to images and Chapter 4.

The Two-Dimensional Fourier Transform

This chapter is about two-dimensional (2-D) functions that change in space, not only horizontally as in 1-D functions but also vertically. We shall focus more on the discrete forms in this chapter, partly because the continuous version is less commonly used but more because images, 2-D sampled functions, pervade our modern lives. There are two academic targets here: *image processing* usually concerns processing images; whereas *computer vision* generally concerns their understanding. Both can employ *feature extraction* to find shapes [Nixon & Aguado, 2019] though vision implies more understanding than processing. Applications/techniques of computer vision include detecting people, understanding 3-D scenes and vision-based biometrics; applications/techniques of image processing include coding, restoration and filtering. One approach for this chapter would be just to process images from the start, but this implies we see exactly what is in front of us. Our human vision is not actually precise and sometimes this can lead to odd effects so we shall consider it first. Images are sampled signals too, so we must consider the sampling process for 2-D signals before moving to 2-D transforms.

4.1 2-D FUNCTIONS AND IMAGES

4.1.1 Image formation

We are familiar with waves – they are part of the sea (or the swimming pool for those inland). These are *2-D continuous waves* varying with space (and time) as in Figure 4.1(a). When we view the waves as a *surface*, well – an approximated one, we have Figure 4.1(b) and when we view this from above we have an *image*, as in Figure 4.1(c). The surface was defined as a (discrete) matrix of points using Matlab, whereas Figure 4.1(a) derives from a camera.

FIGURE 4.1 2-D functions: (a) waves on the sea; (b) 2-D surface; (c) 2-D image

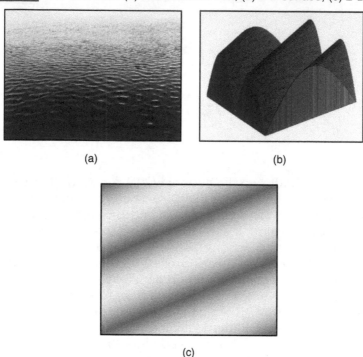

(a)

(b)

(c)

4.1.2 Human vision

Given that we shall view images, we shall briefly look at *human vision* because we do not necessarily perceive the image precisely as formed. There will be effects of sampling, and those shall be covered in the next section. There are other effects, largely introduced by the complexity of the human vision system and we shall cover them briefly to note any *illusions* that might be perceived. Figure 4.2 shows the three main stages in human vision: the eye sensor, neural processing and brain processing, which are modelled (or understood) physically, experimentally and by psychology, respectively. The eye transforms light into neural signals, and the image is formed using a *lens*. The sensors are *cones* or *rods* giving *colour* and *monochrome* vision, respectively (your eyeball is rotated to ensure you make the best use of the current light conditions). This implies that the visual field is sampled spatially on the back of the eye (the retina). The outputs of the sensors give *neural signals,* which are combined, first for transmission down the optic nerve and later via *lateral geniculate nuclei*. The combined signals are passed to the *occipital* and *associative* cortices, which are used to understand the patterns and their links.

FIGURE 4.2 Human vision system

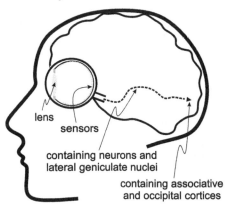

lens

sensors

containing neurons and
lateral geniculate nuclei

containing associative
and occipital cortices

> There is a way of demonstrating that the eye's image is formed upside down, involving a biro and one of your eyelids but we won't go any further as it's quite dangerous – and we won't pay damages!

This is a very complicated system and one which has evolved for many years and works (usually) very well indeed. By physics, light travels in straight lines so the image is formed upside down, so there must be a bit of processing involved. There must also be some form of coding because there are too many sensors for their information to fit within the bandwidth of the optic nerve. Surprisingly, the eye is not particularly good at sensing blue colours and so must compensate for this with visual processing. Brightness adaption can introduce *Mach bands*, in Figure 4.3(a): these are the darker lines between the shaded areas. These do not actually exist but are perceived (introduced by your eye) due to a process known as brightness adaption. The shaded areas in Figure 4.3(a) are simply constant levels of grey and not only do the bands (or stripes) appear, but also some people notice that the grey levels appear to change near the bands. Figure 4.3(b) is ambiguous because it could be a white triangle on three black circles, or three pac-men about to eat each other. There is no information in this image to help us to resolve this.

In Figure 4.3(c) *Benham's disk* is shown as static. When it is cut out and mounted on a disk and then spun, the short black lines do not appear to be grey as one might guess but, amazingly, they appear to have colour (there is a version of this disk on the book's website https://www.southampton.ac.uk/~msn/Doh_Fourier/ so if you have a printed copy of the book you don't need to ruin it). The order of the colours changes when you spin the disk in the opposite direction – wow! The effect appears largely a consequence of the dark side, because if it is removed then only dull grey circles are seen. There are many more

FIGURE 4.3 Illusions in human vision: (a) Mach bands; (b) triangle or pacmen?; (c) Benham's disk

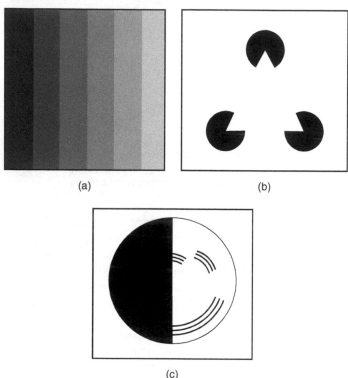

(a)   (b)

(c)

illusions and they have been well studied, particularly in marketing because no one wants the viewer to see the wrong thing, or to be put off by an advert. But the illusions are real: one does not necessarily perceive precisely what has been displayed, and we need to know that when viewing any image.

4.1.3 Sampling images

In Section 3.1.2, the choice of sampling frequency for 1-D signals is dictated by the sampling criterion. There is actually no sampling criterion for 2-D functions (yet). However, we do still have to sample at an appropriate frequency, or signals become aliased. This can be seen in computer images as illustrated in Figure 4.4, which shows an image of a group of people (the computer vision research team at Southampton) displayed at different spatial resolutions (the contrast has been increased

There is potential for confusion in Figure 4.4: if you are reading a version of this book printed at high resolution it should be fine, as written; if you are reading a version that has been printed on a low-resolution printer (or viewing it at low resolution), then the blinds might appear dodgy in a different way. That could rather mince the message here.

FIGURE 4.4 Aliasing in sampled imagery: (a) high resolution; (b) low resolution (aliased)

(a)

(b)

to the same level in each sub-image). Essentially, the people become less distinct in the lower-resolution image, Figure 4.4(b). Now, look closely at the window blinds behind the people. At higher resolution, in Figure 4.4(a), these appear as normal window blinds. In Figure 4.4(b), which is sampled at a much lower resolution, a new pattern appears: the pattern appears to be curved – and if you consider the blinds' relative size, the shapes actually appear to be much larger than normal window blinds. By reducing the resolution, we are seeing an alias of the true information, which appears to be there by a result of sampling. This is *spatial aliasing*. (Remember, we saw this earlier with Mrs. Dali, in Figure 3.3; her alias was President Lincoln.) It is a consequence of sampling whichever way the blinds appear.

As in Section 3.1, to avoid aliasing we must sample at twice the maximum frequency of the signal. There were guidelines for old television signals where a continuous analogue signal from the camera carrying lines of continuous video signals was sampled to provide the digital numbers. The maximum frequency was defined to be 5.5 MHz so we had to sample the camera signal at 11 MHz. Given the timing of a video signal, sampling at 11 MHz implied a minimum image resolution of 576×576 pixels. This was unfortunate: 576 is not an integer power of two and that had poor implications for storage and processing (especially via the FFT). These days we have CMOS and CCD sensors in our cameras, where charge is collected at discrete sites and then converted into digital numbers. As this information is sampled implicitly, we must anticipate the existence of aliasing. This is mitigated somewhat by the observations that:

1. globally, the lower frequencies carry more information, whereas locally the higher frequencies contain more information so the corruption of high-frequency information is of less importance;
2. there is limited *depth of focus* in imaging systems (reducing high-frequency content); and
3. the resolution of sensors used to form digital images continues to increase.

One can still see aliasing when people on television wear checked shirts. It gives rather an ouch to the eyes and they should have known better (certainly if they are presenters).

The *compressive sensing* approach [Donoho, 2016] takes advantage of the fact that many signals have components that are significant, or nearly zero, leading to cameras that acquire significantly fewer elements to represent an image. This provides an alternative basis for compressed image acquisition without loss of resolution.

The effects of *temporal sampling* can often be seen in films, especially in the rotating wheels of cars, as illustrated in Figure 4.5. This shows a wheel with a single spoke, highlighted for simplicity. The film is a sequence of frames starting on the left. The sequence of frames plotted, Figure 4.5(a), is for a wheel

> When watching a film, we bet you haven't thrown a wobbly because the car's going forwards, whereas the wheels say it's going the other way.

that rotates by around 60° between frames, as illustrated in Figure 4.5(b). If the wheel is rotating much faster, by about 300° between frames, as in Figure 4.5(c) and Figure 4.5(d), to a human viewer the wheel will appear to rotate in the opposite direction. This is *temporal aliasing*. If the wheel was exact and rotated by integer multiples of 45°

FIGURE 4.5 Correct and incorrect apparent wheel motion: (a) oversampled rotating wheel; (b) slow rotation; (c) undersampled rotating wheel; (d) fast rotation

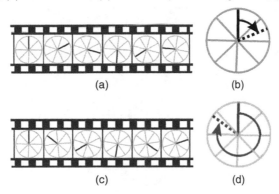

(a)   (b)

(c)   (d)

between frames, it would appear to be stationary. (It's actually more complex than this analysis, but at least it doesn't do a 'Benham's disk' – that would be well scary.) Our eye can resolve this in films because we know that the direction of the car must be consistent with the motion of its wheels, and we expect to see the wheels appear to go the wrong way, sometimes. Curiously, this can also happen with human vision. Does this mean our vision is sampled temporally as well as spatially? (Note that the curious bands sometimes seen in sets of railings are called Moiré fringes and are the result of an interference effect.)

4.1.4 Discrete images

A computer image, Figure 4.1(c), is a matrix (a 2-D array) of *pixels*. The value of each pixel is proportional to the brightness of the corresponding point in the scene; its value is derived from the output of an A/D converter. We can define a square image as $N \times N$ B-bit pixels, where N is the number of points and B controls the number of brightness values. Using B bits in an unsigned integer gives a range of 2^B values, ranging from 0 to $2^B - 1$. If $B = 8$ this gives brightness levels ranging between 0 and 255, which are usually displayed as black and white, respectively, with shades of grey in-between, as they are for the *greyscale image* of a scene in Figure 4.6(a). Smaller values of B give fewer available levels reducing the available contrast in an image. We are concerned with images here, not their formation; image geometry (pinhole cameras etc.) and 3-D analysis can be found elsewhere [Nixon & Aguado, 2019].

The ideal value of the number of bits B is actually related to the *signal-to-noise ratio* (*dynamic range*) of the camera. This is stated as approximately 45 dB for an analogue camera and because there are 6 dB per bit, then 8 bits will cover the available range. Human vision

FIGURE 4.6 Decomposing an image into its bits: (a) original image; (b) bit 0 (LSB); (c) bit 1; (d) bit 2; (e) bit 3; (f) bit 4; (g) bit 5; (h) bit 6; (i) bit 7 (MSB)

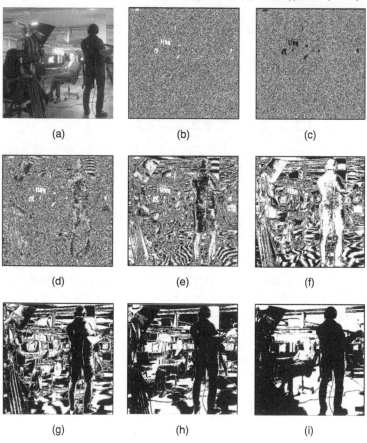

is around 6-7 bits, so we are using more bits in a computer. Choosing 8-bit pixels has further advantages in that it is very convenient to store pixel values as bytes, and 8-bit A/D converters are cheaper than those with a higher resolution. For these reasons, images are nearly always stored as 8-bit bytes, though some applications use a different range. These are the 8-bit numbers for pixels encountered in the Matlab code, stored as unsigned 8-bit bytes (uint8) – anything else is usually double-precision floating point. The relative influence of the 8 bits is shown in the grey-level version of the image of the subjects in Figure 4.6. Here, the least significant bit, bit 0 (Figure 4.6(b)), carries the least information (it changes most rapidly) and is largely noise. As the order of the bits increases, they change less rapidly and carry more information. The most information is carried by the most significant bit, bit 7 (Figure 4.6(i)). Clearly, the fact that there are people in the original image can be recognised much better from the high-order bits, much more reliably than it can from the other bits. There are also odd effects that would appear to come from lighting in the middle order bits. The variation in lighting is hardly perceived from the original image by human vision (but it is there – is this another illusion?).

Colour images specify pixels' intensities in a similar way. However, instead of using just one image plane for monochrome, colour images are represented by three intensity components. These components generally correspond to Red, Green and Blue (the *RGB model*) although there are other colour schemes. For example, the *CMYK colour model* is defined by the components Cyan, Magenta, Yellow and blacK. A pixel's colour can be specified in two ways. First, an integer value can be used, with each pixel, as an index to a table that stores the intensity of each colour component. The index is used to recover the actual colour from the table when the pixel is going to be displayed or processed. In this scheme, the table is known as the image's palette and the display is said to be performed by colour mapping. The main reason for using this colour representation is to reduce memory requirements. That is, we only store a single image plane (i.e., the indices) and the palette. This requires less storage than for the red, green and blue components separately and it makes the hardware cheaper; it can have other advantages, for example. when the image is transmitted. The main disadvantage is that the quality of the image is reduced because only a reduced collection of colours is actually used. An alternative way to represent colour is to use several image planes to store the colour components of each pixel. This scheme is known as true colour and it represents an image more accurately, essentially by considering more colours. The most common format uses 8 bits for each of the three RGB components. These images are known as 24-bit true colour and they can contain $16.777,216$ (2^{24}) different colours. In spite of requiring significantly more memory, the image quality and the continuing reduction in cost of computer memory make this format a good alternative, even for storing the image frames from a video sequence. Here we will process grey-level images only because they contain enough information for our needs. Should the image be originally in colour, we will consider processing its luminance only, often computed in a standard way. In any case, the amount of memory used is always related to the image size. We use colour here to display some of the original images (because a collage of monochrome images can be rather uninspiring). The transforms describe the magnitude and phase of the spatial variation of pixels, so colour is not an innate part of the process at all and the transforms will be displayed as greyscale.

Choosing an appropriate value for the number of points along each image axis, N, is far more complicated. We want N to be sufficiently large to resolve the required level of spatial detail in the image. If N is too small, the image will be coarsely *quantised* (or even aliased): lines will appear to be very 'blocky' and some of the detail will be lost. Larger values of N give more detail but need more storage space and the images will take longer to process because there are more pixels. For example, with reference to the image in Figure 4.6(a), Figure 4.7 shows the effect of taking the image at different resolutions. Figure 4.7(a) is a 64 × 64 image that shows only the broad structure. It is impossible to see any detail in the sitting subject's face, or anywhere else. Figure 4.7(b) is a 128 × 128 image, which is starting to show more of the detail, but it would be hard to determine the subject's identity. The image in Figure 4.7(c) is a 256 × 256 image, which shows a much greater level of detail, and the subject can perhaps be recognised from the image. Note that the images in Figure 4.7 have been scaled to be the same size. As such, the pixels in Figure 4.7(a) are much larger, leading to blocky structure, than those in Figure 4.7(c). Common choices are for, say, 512 × 512 or 1024 × 1024 8-bit images, which require 256 KB and 1 MB of storage, respectively. If we take a sequence of, say, 20 images for motion analysis, without coding we need more than 5 MB to store twenty 512 × 512 images. Even though memory continues to become cheaper, this can still impose high cost.

FIGURE 4.7 Effects of differing image resolution: (a) 64 × 64; (b) 128 × 128; (c) 256 × 256

(a)   (b)

(c)

FIGURE 4.8   Image axis conventions and spatial frequency

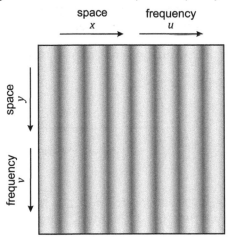

4.1.5 Discrete image frequency components

The axis conventions in images are that *x* is for horizontal and *y* is for vertical. These are the rows and columns of the image matrix, respectively. In software, the image is usually addressed as image(y,x) for reasons of efficient storage access. The *spatial frequency* (the 2-D frequency) is usually denoted *u* for horizontal and *v* for vertical. Figure 4.8 shows the axis conventions and a horizontal spatial frequency. In this image, the vertical frequency *v* = 0, because there is no change in that direction.

4.2 2-D FOURIER TRANSFORM AND ITS INVERSE

4.2.1 2-D continuous Fourier transform and separability

The *2-D continuous Fourier transform* is defined as an extension of the 1-D transform as

$$FP(u, v) = \int_{-\infty}^{\infty} \int_{-\infty}^{\infty} P(x, y)\, e^{-j2\pi(ux+vy)} dxdy \qquad (4.1)$$

following the axis conventions in Figure 4.8. The *inverse 2-D continuous Fourier transform* reconstructs the image and is defined as

$$P(x, y) = \frac{1}{4\pi^2} \int_{-\infty}^{\infty} \int_{-\infty}^{\infty} FP(u, v)\, e^{j2\pi(ux+vy)} dudv \qquad (4.2)$$

Now the two axes are independent and knowledge about the information along one axis implies nothing about the information in the other, so they are known as *orthogonal*. This is evident in Figure 4.8

where the change in one axis is not reciprocated by a change in the other. This is known as *separability* and implies that we can compute the integrals independently of each other as

$$FP(u, v) = \text{vertical transform}\,(\text{horizontal transform}\,(P(x, y)))$$

which as the Fourier equation is

$$FP(u, v) = \int_{-\infty}^{\infty} \overbrace{\left(\int_{-\infty}^{\infty} P(x, y)\, e^{-j2\pi ux} dx \right)}^{\text{horizontal 1-D transform}} e^{-j2\pi vy} dy \qquad (4.3)$$

where the inner term is the transform of the horizontal information, the 1-D transform we first met in Chapter 2, and the vertical transform embraces it.

We shall only dwell here on the continuous 2-D Fourier transform to show separability in action, using angular frequencies $u' = 2\pi u$ and $v' = 2\pi v$, and we shall return to it later in applications, Chapter 6. A 2-D $T \times T$ continuous pulse of magnitude A is defined as

$$P(x, y) = \begin{cases} A & -\dfrac{T}{2} \le x; y \le \dfrac{T}{2} \\ 0 & \text{otherwise} \end{cases} \qquad (4.4)$$

and its Fourier transform is

$$FP(u', v') = \int_{-T/2}^{T/2} \int_{-T/2}^{T/2} A e^{-j(u'x + v'y)}\, dx\, dy$$

and by separability this is

$$FP(u', v') = A \int_{-T/2}^{T/2} \left(\int_{-T/2}^{T/2} e^{-ju'x} dx \right) e^{-jv'y} dy$$

by Equation 2.4

$$FP(u', v') = A \int_{-T/2}^{T/2} \left(\frac{e^{\frac{-ju'T}{2}} - e^{\frac{ju'T}{2}}}{-ju'} \right) e^{-jv'y} dy$$

by rearrangement

$$FP(u', v') = A \left(\frac{e^{\frac{-ju'T}{2}} - e^{\frac{ju'T}{2}}}{-ju'} \right) \int_{-T/2}^{T/2} e^{-jv'y} dy$$

and so the 2-D (sinc) function is

$$FP(u', v') = A \left(\frac{e^{\frac{-ju'T}{2}} - e^{\frac{ju'T}{2}}}{ju'} \right) \left(\frac{e^{\frac{-jv'T}{2}} - e^{\frac{jv'T}{2}}}{jv'} \right) \qquad (4.5)$$

which shows separability in the frequency domain because there are two separate functions in *u* and *v*. This is a 2-D transform pair, and these naturally exist for other functions.

4.2.2 2-D discrete Fourier transform

For a square $N \times N$ image the *2-D DFT* is given in terms of the horizontal *u* and vertical *v* frequency components

$$FP_{u,v} = \frac{1}{N^2} \sum_{x=0}^{N-1} \sum_{y=0}^{N-1} P_{x,y} e^{-j\frac{2\pi}{N}(ux+vy)} \qquad u; v = 0, 1 \ldots N-1 \qquad (4.6)$$

where the scaling coefficient $\frac{1}{N^2}$ ensures that the d.c. term $FP_{0,0}$ is the average of the points in the image **P**. Equation 4.6 is the definition we shall use from now on though there are other definitions as discussed in Section 7.1.1. The DFT can be calculated using separability as

$$FP_{u,v} = \frac{1}{N^2} \sum_{y=0}^{N-1} \left(\sum_{x=0}^{N-1} P_{x,y} e^{-j\frac{2\pi}{N}ux} \right) e^{-j\frac{2\pi}{N}vy}$$

and thus by FFTs along each axis

$$\mathbf{FP} = \text{FFT}y\,(\text{FFT}x\,(\mathbf{P})) \qquad (4.7)$$

Because practical images can contain more than 10^6 points, Equation 4.7 shall be used from now on (in Matlab it is `fft2` because `fft` is a 1-D operation only, though it could be invoked in succession deploying separability). Some basic image frequencies and their transforms are shown in Figure 4.9. We shall centre the transforms for display so that the d.c. component is at the centre, and scale the brightness; we describe image display in more detail in Section 4.3.1. When the change is only along the horizontal axis and not down the vertical axis, Figure 4.9(a), we have horizontal frequency components, Figure 4.9(b). When the image is rotated, the frequency components are along the vertical axis, Figure 4.9(d), because that is where the change is.

FIGURE 4.9 Basic image frequency components: (a) horizontal change; (b) horizontal frequencies; (c) vertical change; (d) vertical frequencies

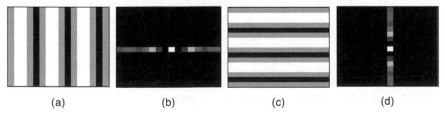

    (a)   (b)   (c)   (d)

FIGURE 4.10 Fourier transform: (a) original image; (b) magnitude; (c) phase

(a)

(b)   (c)

The centred transform of Figure 4.10(a) shows frequency components, Figure 4.10(b), that exist all over the spectrum, particularly those associated with the leaves (the change is normal to the leaves, giving the dominant diagonal axis in the transform). The brightest points (those frequencies with the largest amplitude) are near the lower frequencies, which is common in many practical images. There appears to be little structure in the phase, Figure 4.10(c), except that associated with the leaves and that is implicit in the process and because it is rather difficult to display.

A Matlab implementation of the 2-D DFT is given in Code 4.1, but note that it is too slow for anything except a tiny image (or a supercomputer).

Code 4.1 Discrete Fourier transform in Matlab

```
function [Fourier] = F_transform(image)

image=double(image);
[rows, cols] = size(image);
%we deploy equation 4.9, to handle non-square images
for u=1:cols %along the horizontal axis
  for v=1:rows %down the vertical axis
```

```
    sumx=0;
    for x=1:cols
        %first we transform the rows
        sumy=0;
        for y=1:rows %Eq 4.7 inner bracket
            sumy=sumy+image(y,x)*exp(-1j*2*pi*(v-1)*(y-1)/rows);
        end
        %then we do the columns Eq 4.7 outer
        sumx=sumx+sumy*exp(-1j*2*pi*(u-1)*(x-1)/cols);
    end %and finally normalise
    Fourier(v,u) = sumx/(rows*cols);
end
end
```

The *2-D inverse DFT* of a square image is

$$P_{x,y} = \sum_{u=0}^{N-1} \sum_{v=0}^{N-1} FP_{u,v} e^{j\frac{2\pi}{N}(ux+vy)} \qquad x; y = 0, 1 \ldots N - 1 \tag{4.8}$$

We shall use the inverse DFT to reconstruct an image using components selected from the transform of Figure 4.6(a). The contribution of different frequencies is illustrated in Figure 4.11, where we have reconstruction from a circular region centred on d.c. and of different radii (by centring, $FP_{0,0}$ is in the middle of the transform image). Figure 4.11 shows in the first row (a)–(d) the position of the image transform components within a circle of chosen radius (shown as the logarithm of their magnitude). Below, there is also the image of the magnitude of the Fourier transform, (m). The second row (e)–(h) shows the image reconstructed from frequency components at the chosen radius (the points on the perimeter of the circle). The second row shows some of the component spatial frequencies. The third row (i)–(l) shows the reconstruction (by the inverse FT) using frequency components up to and including the components at that radius, from the transforms in the first row (the points within the circle). In Figure 4.11 the first column is the transform components at radius 4 (which are low-frequency components), the second column is the radius 16 components, the third column is at radius 32 and the fourth column is the radius 64 components (the higher-frequency components). The last row has the magnitude of the Fourier transform image, (m), and the reconstruction of the image from the whole transform, (n). As we include more components we include more detail; the lower-order components carry the bulk of the information. In the third row, the first components plus the

FIGURE 4.11 Image reconstruction and different frequency components:
(a) transform ≤ radius 4 components; (b) transform ≤ radius 16 components;
(c) transform ≤ radius 32 components; (d) transform ≤ radius 64 components;
(e) image by radius = 4 components; (f) image by radius = 16 components;
(g) image by radius = 32 components; (h) image by radius = 64 components;
(i) reconstruction from (a); (j) reconstruction from (b); (k) reconstruction from (c);
(l) reconstruction from (d); (m) complete transform; (n) reconstruction from (m)

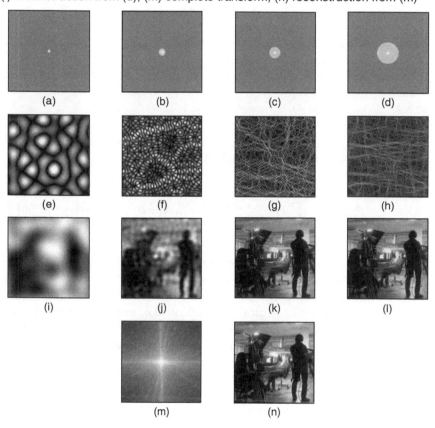

d.c. component give a very coarse approximation (i); when the components up to radius 16 are added we can see the shape of the people (j); the components up to radius 32 allow us to see features (k), but they are not sharp; we can recognise items from the components up to radius 64 (l); when all components are added (n) we return to the original image. This also illustrates *coding*, as the image can be encoded by retaining fewer of the components of the image than are in the complete transform. Figure 4.11(l) (and even Figure 4.11(k)) is a good example of where an image of acceptable quality can be reconstructed, even when most of the frequency components have been discarded.

4.3 PROPERTIES OF THE 2-D DISCRETE FOURIER TRANSFORM

4.3.1 Displaying images

4.3.1.1 Transforms and their repetition

One of the important properties of the FT is *replication*, which implies that the transform repeats in frequency up to infinity, as indicated in Figure 3.5 for 1-D signals. To show this for 2-D signals, we need to investigate the Fourier transform, here for non-square $M \times N$ images

$$FP_{u,v} = \frac{1}{MN} \sum_{x=0}^{N-1} \sum_{y=0}^{M-1} P_{x,y} e^{-j\left(\frac{2\pi}{N} ux + \frac{2\pi}{M} vy\right)} \tag{4.9}$$

The Fourier transform at integer multiples of the number of sampled points $FP_{u+nN,v+mM}$ (where m and n are integers) is, by substitution in Equation 4.9:

$$FP_{u+nN,v+mM} = \frac{1}{MN} \sum_{x=0}^{N-1} \sum_{y=0}^{M-1} P_{x,y} e^{-j\left(\frac{2\pi}{N}(u+nN)x + \frac{2\pi}{M}(v+mM)y\right)}$$

$$FP_{u+nN,v+mM} = \frac{1}{MN} \sum_{x=0}^{N-1} \sum_{y=0}^{M-1} P_{x,y} e^{-j\left(\frac{2\pi}{N} ux + \frac{2\pi}{M} vy\right)} \times e^{-j2\pi(nx+my)}$$

and because $e^{-j2\pi(nx+my)} = 1$ (because the term in brackets is always an integer, the exponent is always an integer multiple of 2π),

$$FP_{u+nN,v+mM} = \frac{1}{MN} \sum_{x=0}^{N-1} \sum_{y=0}^{M-1} P_{x,y} e^{-j\left(\frac{2\pi}{N} ux + \frac{2\pi}{M} vy\right)} = FP_{u,v} \tag{4.10}$$

which shows that the replication property does indeed hold for the 2-D DFT.

To display centred transforms of images (as we have indeed been doing so far), we use the same operation as for the 1-D DFT (Section 3.2.3) and here over two dimensions

$$FP_{u,v} = \frac{1}{MN} \sum_{x=0}^{N-1} \sum_{y=0}^{M-1} P_{x,y} e^{-j\left(\frac{2\pi}{N} ux + \frac{2\pi}{M} vy\right)} \times -1^{x+y} \tag{4.11}$$

2-D replication is shown in Figure 4.12. The display operation means that we are shifting the origin of the transform so that d.c. is at

FIGURE 4.12 Transform repetition and centring the 2-D transform

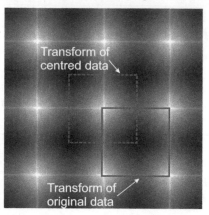

the centre. If transforms are displayed without centring, then the low-frequency components are in the corners and the high-frequency components are in the centre, which does not appeal to intuition. When transforms are centred it is the other way round and the low frequencies are those nearest the middle. We shall revisit this when we apply the 2-D Fourier transform, particularly its use in convolution.

As with the display of 1-D transforms (Section 3.2.3), it is common to use logarithms of magnitude to display the Fourier transforms of images. This is particularly the case for 2-D transforms, as the d.c. coefficient can be very much larger than the other components, making the transform difficult to see (as there is a bright point in the middle, and nowt else).

4.3.1.2 Intensity normalisation

In Section 4.1.4 we described how the 8-bit integers often used to store images lead to 256 brightness values between 0 and 255. When we process images, it is often as floating-point numbers, which are converted for display. To ensure that an image is visible, we can use *intensity normalisation* to display images using values between 0 and 255 as

$$\text{normalised\_}\mathbf{P} = (\mathbf{P} - \min(\mathbf{P})) \times \frac{255}{\max(\mathbf{P}) - \min(\mathbf{P})} \quad (4.12)$$

which for a combination of $\max(\mathbf{P}) = 255$ and $\min(\mathbf{P}) = 0$ gives no scaling at all, as required. This is Matlab's `imagesc` function, which displays images nicely. There is a different approach for display called *histogram equalisation*, which scales brightness in a manner aimed to be suited to human vision, but it is highly nonlinear and not used here (and loathed in automated image analysis for that reason).

4.3.2 Rotation

The Fourier transform of an image rotates when the source image rotates. This is to be expected because the decomposition into spatial frequency reflects the orientation of features within the image. As such, orientation dependency is built into the Fourier transform process. This implies that if the frequency domain properties are to be used in image analysis, via the Fourier transform, the orientation of the original image needs to be known, or fixed (or even, using prior knowledge, estimated from the transform itself). It is often possible to fix orientation or to estimate its value when a feature's orientation cannot be fixed. Alternatively, there are techniques to impose invariance to rotation, say by translation to a polar representation, though this can prove to be complex. For a rotation R of an image **P** by an angle θ, the transform rotates by the same amount as the image

$$\mathcal{F}\left(R\left(\mathbf{P};\theta\right)\right) = R\left(\mathbf{FP};\theta\right) \tag{4.13}$$

For analysis we shall use a *rotation matrix,* which rotates points with coordinates (x, y) by an angle θ to new coordinates (x', y') as

$$\begin{bmatrix} x' \\ y' \end{bmatrix} = \begin{bmatrix} \cos\theta & -\sin\theta \\ \sin\theta & \cos\theta \end{bmatrix} \begin{bmatrix} x \\ y \end{bmatrix} \tag{4.14}$$

The rotated transform is

$$\mathcal{F}\left(R\left(\mathbf{P};\theta\right)\right) = \mathcal{F}\left(P_{x',y'}\right)$$

$$= \frac{1}{MN} \sum_{x'=0}^{N-1} \sum_{y'=0}^{M-1} P_{x',y'} e^{-j\left(\frac{2\pi}{N}u'x' + \frac{2\pi}{M}v'y'\right)}$$

From Equation 4.14

$$= \frac{1}{MN} \sum_{x=0}^{N-1} \sum_{y=0}^{M-1} P_{x,y} e^{-j\left(\frac{2\pi}{N}u'(x\cos\theta - y\sin\theta) + \frac{2\pi}{M}v'(x\sin\theta + y\cos\theta)\right)}$$

By collecting terms in x and y

$$= \frac{1}{MN} \sum_{x=0}^{N-1} \sum_{y=0}^{M-1} P_{x,y} e^{-j\left(\frac{2\pi}{N}x(u'\cos\theta + v'\sin\theta) + \frac{2\pi}{M}y(v'\cos\theta - u'\sin\theta)\right)}$$

The transform's rotation within this is $\begin{bmatrix} u \\ v \end{bmatrix} = \begin{bmatrix} \cos\theta & \sin\theta \\ -\sin\theta & \cos\theta \end{bmatrix} \begin{bmatrix} u' \\ v' \end{bmatrix}$ and by inversion

$$\begin{bmatrix} u' \\ v' \end{bmatrix} = \begin{bmatrix} \cos\theta & -\sin\theta \\ \sin\theta & \cos\theta \end{bmatrix} \begin{bmatrix} u \\ v \end{bmatrix} \tag{4.15}$$

and the rotated transform has the same rotation as for the image.

The effect of rotation is illustrated in Figure 4.13, on a picture of some rather nice scarves in Myanmar (Mark bought one too). An image, Figure 4.13(a), is rotated to give the image in Figure 4.13(c) (noting again that colour plays no part in the transform process, it is just for display). Comparison of the transform of the original image, Figure 4.13(b), with the transform of the rotated image, Figure 4.13(d)

FIGURE 4.13 Illustrating rotation: (a) original image; (b) transform of original image; (c) rotated image; (d) transform of rotated image

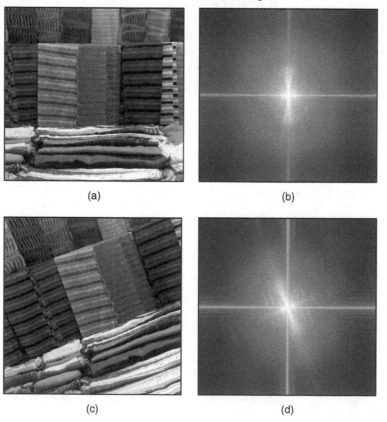

(a)   (b)

(c)   (d)

shows that the transform has been rotated by the same amount as the image. In fact, close inspection of Figures 4.13(b) and (d) shows that the diagonal axis is consistent with the normal to the axis of bars of the cloth (where the change mainly occurs), and this is the axis that rotates.

4.3.3 Scaling

The scaling property associated with images arises from imaging at different distances, as illustrated in Figure 4.14 where the image of spots (a 3-D calibration target, which is used to determine the camera parameters for 3-D reconstruction), Figure 4.14(a), is reduced in scale, Figure 4.14(b), thereby reducing the spatial frequency (both images were sampled from an original image and have the same number of points). The DFT of the original image is shown in Figure 4.14(c), which reveals that the large spatial frequencies in the original image are arranged in a galaxy-like pattern. The inverse relationship between space and frequency maintains (as shown for 1-D signals in Section 3.3.4): as a consequence of effectively stretching the original image, the spectrum will appear to have lower spatial frequency, as shown in Figure 4.14(d). This retains the pattern of the ring-like structure. The relationship is linear: the amount of reduction, say the proximity of the camera to the target, is directly proportional to the scaling in the frequency domain.

FIGURE 4.14 Illustrating frequency scaling: (a) 3-D target; (b) scaled 3-D target; (c) transform of 3-D target; (d) transform of scaled 3-D target

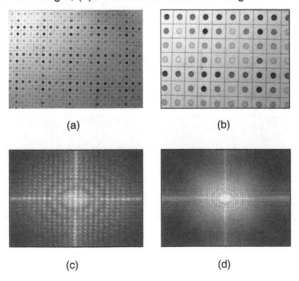

(a)   (b)

(c)   (d)

4.3.4 Shift invariance

The 2-D DFT also has shift invariance like its 1-D version, and this is simply a consequence of shifting the position of the reconstructed transform, as in Figure 4.12. This process is illustrated in Figure 4.15. An original image (of tourists on a camel ride in the desert in Morocco), Figure 4.15(a), is shifted along the x- and the y-axes, Figure 4.15(d). The shift is cyclical, so parts of the image wrap around; those parts at

FIGURE 4.15 Illustrating shift invariance: (a) original image; (b) magnitude of Fourier transform of original image; (c) phase of Fourier transform of original image; (d) shifted image; (e) magnitude of Fourier transform of shifted image; (f) phase of Fourier transform of shifted image

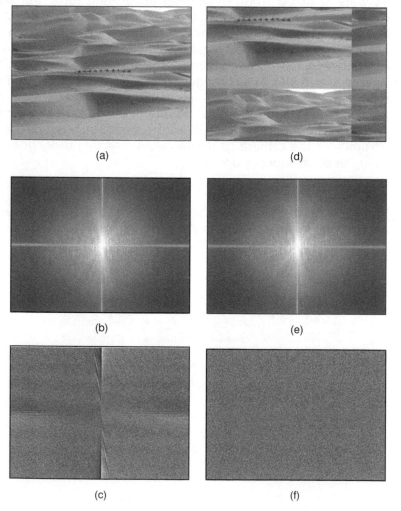

(a) (d)

(b) (e)

(c) (f)

the top of the original image appear at the base of the shifted image. The shifted image appears nonsensical. The magnitude of the Fourier transform of the original image and of the shifted image is identical: Figure 4.15(b) appears the same as Figure 4.15(e). The phase differs: the phase of the original image, Figure 4.15(c), is clearly different from the phase of the shifted image, Figure 4.15(f). This is dramatic in images: given that Figure 4.15(d) is simply a mess (the skyline is now in the middle of the image, and the dunes are discontinuous), it does appear surprising that the magnitude of the resulting transform does not change. We are, however, seeing the effects of sampled image/transform replication, as shown in Figure 4.12.

The shift invariance property implies that, in application, the magnitude of the Fourier transform of a texture, say, will be the same irrespective of the position of the texture in the image (i.e. the camera or the target can move), assuming that the texture is much larger than its imaged version. This implies that if the magnitude of the Fourier transform is used to analyse an image of a human face or an image of some cloth, to describe it by its spatial frequency, we do not need to control the position of the camera, or the cloth, because the magnitude of the transform components will be unchanged.

4.3.5 The importance of phase

The Fourier transform has been used in more biometrics than gait, including: fingerprints; palm; vein and ECG. Phase is especially important in iris biometrics.

The importance of phase is much clearer with images than with 1-D functions (e.g. Section 2.4). An early study [Oppeheim81] showed that 'an image can be exactly recovered from its phase function to within a scaling factor if the image satisfies the appropriate conditions'. We shall take two images, one of an eye and the other of an ear, as shown in Figure 4.16. Here we have reconstructed images by taking the magnitude of one image and the phase (the argument) of another. From Equation 2.6 we have

$$FP_{u,v} = \text{Re}\left(FP_{u,v}\right) + j\,\text{Im}\left(FP_{u,v}\right) \tag{4.16}$$

where magnitude and phase are calculated according to Equations 2.7 and 2.8, respectively. The rearranged Fourier transform is then

$$P_{x,y} = \mathcal{F}^{-1}\left(|FP1_{u,v}| \times \cos\left(\arg\left(FP2_{u,v}\right)\right) + j \times |FP1_{u,v}| \times \sin\left(\arg\left(FP2_{u,v}\right)\right)\right) \tag{4.17}$$

When image 1 is the eye and image 2 is the ear, then the reconstructed image (via the inverse FT) looks most like the image of the ear, Figure 4.16(c); when it is the other way found and image 1 is the ear and image 2 is the eye, the resulting image is much closer to the eye,

FIGURE 4.16 Illustrating the importance of phase: (a) eye image; (b) ear image; (c) reconstruction from *magnitude* (eye) and *phase* (ear); (d) reconstruction from *magnitude* (ear) and *phase* (eye)

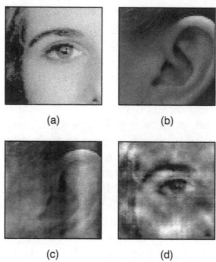

(a)   (b)

(c)   (d)

Figure 4.16(d). It is thus the phase that largely controls the reconstruction in this case, not the magnitude (the bits represented by the magnitude hardly show in the reconstructions here). Clearly the phase is very important in the representation of a signal, and this is much clearer than for the 1-D form (Figure 2.15). In the applications chapter, we shall find an image-processing operator that is predicated entirely by the concept of phase, Section 6.5. Phase is important, get it!

Other properties are covered in applications in Chapter 6 – especially superposition and where it can and cannot be used, in Section 6.3.1.

4.3.6 Computational cost of 2-D DFT and FFT

For N points, the computational cost of the 1-D DFT was $O\left(N^2\right)$ and the 1-D FFT was $O\left(N\log_2 N\right)$. For an $N \times N$ image we have N^2 points, and the operations use separability in that the 2-D versions transform the N point rows as N point columns, so

$$\text{the computational cost of the 2-D DFT} = O\left(N^3\right) \qquad (4.18)$$

$$\text{the computational cost of the 2-D FFT} = O\left(2N^2\log_2 N\right) \qquad (4.19)$$

It is thus completely impractical to perform the DFT on image data, because it just takes too long. Given that the FFT gives the same result with the advantage stated in key point 7, it is difficult to see any reason to use the DFT for images.

4.4 IMAGE PROCESSING VIA THE FOURIER TRANSFORM

4.4.1 Convolution

4.4.1.1 Image convolution

Image convolution, 2-D convolution, is the same form as 1-D convolution but with images rather than signals. In Equation 3.41 1-D circular convolution was given as,

$$\mathbf{conv} = \mathbf{p}\,[m] * \mathbf{q}\,[m] = \sum_{x=0}^{C-1} \mathbf{p}\,[x]\,\mathbf{q}\,[m-x]$$

where C is the number of points (or columns) in the vectors $\mathbf{p}$ and $\mathbf{q}$. This involved multiplication by a flipped version of $\mathbf{q}$. Our signals are now images so we have two dimensions $R \times C$ and we invert (or flip) the image $\mathbf{Q}$ around both axes.

$$\mathbf{CONV} = \mathbf{P} * \mathbf{Q} = \sum_{x=0}^{C-1}\sum_{y=0}^{R-1} \mathbf{P}\,[x,y]\,\mathbf{Q}\,[m-x, n-y] \tag{4.20}$$

$$m = 0, 1 \ldots C - 1; n = 0, 1 \ldots R - 1$$

This is compact notation indeed. The direct implementation of circular image convolution is

$$CONV_{x,y} = \sum_{x=0}^{C-1}\sum_{y=0}^{R-1} P_{x,y} \times Q_{\mathrm{mod}(x-m,C-1),\mathrm{mod}(y-n,R-1)} \tag{4.21}$$

$$m \in 0, 1 \ldots C - 1, R; n \in 0, 1 \ldots R - 1$$

where $\mathrm{mod}\,()$ is again the modulo process (as in Equation 3.42). As we found in Section 3.4 the Fourier transform can be used to speed up the convolution process. Image convolution is a 2-D version of Equation 3.45.

$$\mathbf{P} * \mathbf{Q} = \mathcal{F}^{-1}\,(\mathcal{F}\,(\mathbf{P}) . \times \mathcal{F}\,(\mathbf{Q})) \tag{4.22}$$

and because the amount of computation in the basic form is $O\left(C^2 \times R^2\right)$, the Fourier implementation is invariably used (to be considered in detail in Section 4.4.2). With the Fourier transform the transforms of the images need to be the same size before we can perform the point-by-point multiplication. Accordingly, zero-padding can be used, which simply means that points that are zero are added to make the images the same size.

4.4.1.2 Template convolution

Image convolution is rarely used in image processing/computer vision where the process of *template convolution*, as used in many operators, calculates points by convolving an image with a *template* **w** where the template of size $C_T \times R_T$ is (much) smaller than the image with which it is convolved, as

$$T\_CONV_{x',y'} = \sum_{x=0}^{C_T-1} \sum_{y=0}^{R_T-1} P_{x'-\frac{C_T}{2}+x,y'-\frac{R_T}{2}+y} \times w_{C_T-1-x,R_T-1-y}$$

$$\begin{bmatrix} x' \in C_T/2, C-1-C_T/2 \\ y' \in R_T/2, R-1-R_T/2 \end{bmatrix} \qquad (4.23)$$

where $x' - \frac{C_T}{2} + x, y' - \frac{R_T}{2} + y$ are the coordinates of the point that is multiplied by the template point w, as shown in Figure 4.17. As the figure shows, the (flipped) template is convolved with the image in a raster fashion, scanning from left to right, top to bottom. The size of the template implies that the final image is smaller than the original image, as should the template extend beyond the image there would be no information to process. In direct implementations of template convolution, a border (usually black = 0) is added so the output image is the same size as the original image. Alternatives to adding a border naturally include the use of modulo arithmetic (invoking replication in the space domain) or using a smaller template at the borders of an image but these are rarely used, largely because the templates are often quite small (the border also adds a rather neat frame to the

FIGURE 4.17 Illustrating direct implementation of template convolution: (a) process; (b) result

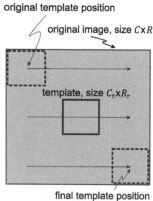
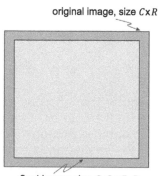

(a)   (b)

image). Circular convolution is generally not employed in direct implementations of template though convolution can be achieved by using the convolution theorem with a zero-padded template.

The result of template convolution depends on the values used in the elements of the template, and this leads to the many different operators in computer vision. (Note that convolutional neural networks actually perform correlation via symmetric templates, which we shall cover in the next section.) If each element of the template $w_{i,j}$ is set to 1.0, then the result of template convolution is the addition of all points covered by the template. If this sum is divided by the area of the template (the number of its points), then the result is averaging. This can then be denoted as the summation of pixel values inside the template

$$Av_{x',y'} = \frac{1}{C_T \times R_T} \sum_{x=0}^{C_T-1} \sum_{y=0}^{R_T-1} P_{x'-\frac{C_T}{2}+x, y'-\frac{R_T}{2}+y} \quad \begin{bmatrix} x' \in C_T/2, C-1-C_T/2 \\ y' \in R_T/2, R-1-R_T/2 \end{bmatrix}$$

$$(4.24)$$

We shall show the application of this operator to an image in Figure 4.18, using a template size 15×15. The template is a larger version (and differently scaled) of the template in Figure 4.18(c) and the scaling is usually applied to the result rather than when convolving the template, for speed. The template size is usually an odd number and in direct implementations the result is stored at the position of the centre point of the template in the new image. The result in Figure 4.18 is blurred, as expected by the averaging process, and there is a clear window where points were not computed. The template size is chosen so we can actually see the border; if we had used a 3×3 template, we would not be able to see the border at all.

Just having an operator that blurs images might appear rather useless (soft focus for romantic effect is not much use here) though small

FIGURE 4.18 Averaging an image by template convolution: (a) image of lizard; (b) averaged image; (c) template for 3×3 operator

1/9	1/9	1/9
1/9	1/9	1/9
1/9	1/9	1/9

(a)   (b)   (c)

FIGURE 4.19 Averaging a noisy image by template convolution: (a) noisy image; (b) averaged image; (c) noisy image (detail); (d) averaged image (detail)

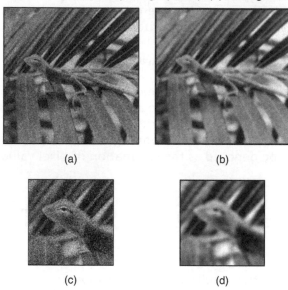

(a)   (b)

(c)   (d)

aperture cameras in smartphones can use out-of-focus blurring to syn-thesise shallow depth of field imagery [Seely et al., 2019]. Here, we shall stay simple and use the averaging operator to remove noise. We shall add some noise to the image of the lizard in Figure 4.18(a), shown in Figure 4.19(a) (and to show the noise, there is an image of some detail – the lovely lizard's head – in Figure 4.19(c)). The result of averag-ing with an 11×11 averaging operator cleans the image, Figure 4.19(b), and the detail in Figure 4.19(d) shows that you do not get something for nothing because the resulting image has less noise, but more blur-ring. The effect of smoothing is to reduce noise but to blur the image. There are much more sophisticated filtering/restoration approaches available, which we shall start to cover in Section 4.4.4 and later those approaches, which aim to preserve detail but reduce noise, such as non-local means and bilateral filtering, in Section 6.6. All of these can provide a superior performance to averaging. For now, we are concen-trating on convolution itself, so let us speed up the process by using Fourier.

4.4.1.3 Filtering an image via convolution

The convolution theorem, Equation 4.22, is used for large tem-plates, and a direct implementation can give a faster result for small templates as we shall see in the next section. Averaging by image convolution is illustrated in Figure 4.20. This is implemented using the

FIGURE 4.20 Template/image convolution via Fourier transform: (a) image of lizard; (b) zero-padded averaging template; (c) resulting averaged image; (d) image transform; (e) template transform; (f) multiplied transforms

(a)   (d)

(b)   (e)

(c)   (f)

FFT, Equation 4.22, though the same result would be achieved by template convolution. The image of an averaging template, Figure 4.20(b), is convolved with an image, Figure 4.20(a), using Equation 4.22. The averaging template has been zero-padded to be the same size as the image, Figure 4.20(b). The Fourier transform of the template, Figure 4.20(e), is multiplied by the Fourier transform of the image, Figure 4.20(d), to form the transform in Figure 4.20(f). When viewing the combined-transform image compared with those it is derived from, note that there are some odd brightness effects due to normal-isation. The multiplication provides a transform, Figure 4.20(f), which is then inverse Fourier transformed to give the smoothed image in Figure 4.20(c). As before, the averaging process smooths the image, reducing change and also blurring the image features. The (2-D FFT) transforms are shown as centred and as the logarithm of their magni-tude. The transform of the averaging template is a 2-D sinc function viewed from above. The template here was 15×15 and the image res-olution is around 800×800 so the template size is chosen with this in mind – and so we can see it.

If a direct implementation had been used, the resulting image would be smaller (or the same size with a black border, as in Figure 4.21(a)). The transform-based approach has resulted in an image that is the same size, without a border, Figure 4.21(c). This is because the transform-based implementation assumes that the image replicates and so there is some information in the border though it appears to be incorrect in Figure 4.20(c) because the image is of course actually a sample from a much larger view, and one that is not repli-cated. The difference between the two, Figure 4.21(c), shows the dif-ference in the border regions very clearly, and that there is (very) little other difference between the averaged images.

4.4.2   Computational considerations of image convolution and template convolution

Templates are generally smaller than the image size so there is a chance that for smaller size templates the Fourier transform offers no com-putational advantage, because the time spent processing the padded points exceeds the time spent processing the template points. The question to be answered here is how small should the template be before it is advantageous to use the direct implementation of template convolution, Equation 4.23, rather than the Fourier implementation of image convolution, Equation 4.22.

For simplicity, we shall consider square images and templates. For a square image with N^2 points, the computational cost of a 2-D FFT, from Equation 4.19, is of the order of $2N^2 \log_2 N$. Given a precomputed transform of the template, in application to an image, there are two transforms required and there is one multiplication for each of the $N \times N$ transformed points. The total cost of the Fourier implementation of template convolution is then

$$\text{Cost}_{\text{FFT}} = O\left(4N^2 \log_2 N + N^2\right)$$

The cost of the direct implementation of template convolution for a square $m \times m$ template is m^2 multiplications for each image point, so

FIGURE 4.21 Difference between direct and Fourier implementations of template convolution (31 × 31 operator): (a) direct (template) averaging; (b) averaging via convolution theorem; (c) difference (a)-(b)

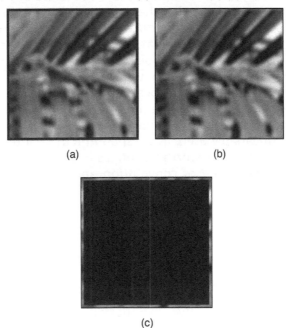

(a)   (b)

(c)

the cost of the direct implementation of template convolution is of the order of

$$\text{Cost}_{\text{direct}} = O\left(N^2 m^2\right)$$

For $\text{Cost}_{\text{direct}} < \text{Cost}_{\text{FFT}}$, we require:

$$N^2 m^2 < 4N^2 \log_2 N + N^2$$

If the direct implementation of template convolution is to be faster than its Fourier implementation, we need to choose m so that

$$m^2 < 4 \log_2 N + 1 \qquad (4.25)$$

This implies that, for a 256 × 256 image, a direct implementation is fastest for 3 × 3 and 5 × 5 templates, whereas a transform-based calculation is faster for larger ones. For a 4096 × 4096 image it is also just faster to convolve a 7 × 7 template directly. In OpenCV (a software package often used to implement computer vision algorithms), for some versions the limit appears to be 7 × 7 [OpenCV-TM] for using the FFT and for a kernel size of 5 × 5 or less a direct version is used which suggests the same limit as the analysis here. There are higher considerations of complexity than our analysis has included, as mentioned in Section 3.6.4, and the complexity and compatibility issues preclude investigation of Equation 4.25 via the Matlab implementations (though the cost of the direct implementation is easy to verify). There are, naturally, further considerations in the use of transform calculus, the most important being the use of windowing (such as Hamming or Hanning) operators to reduce variance in high-order spectral estimates. This implies that template convolution by transform calculus should be used when large templates are involved, and when speed is critical. Image convolution uses the transform-based implementation only. Templates in image processing are often small for the less complex operators motivating the use of Equation 4.23, but for some operators the templates are routinely larger than 7×7 when Fourier convolution is invariably used, as we shall find in Chapter 6 when we consider applications. It is worth noting (as in Section 3.6.4) that deep learning makes much use of convolution and especially for 2-D signals there is interest [Gu et al., 2018; Lavin & Gray, 2016] in the fastest approaches, which are now using *Winograd convolution* to improve performance.

4.4.3 Correlation

4.4.3.1 Image correlation

The 1-D form of correlation was in Equation 3.47

$$\mathbf{p}[m] \otimes \mathbf{q}[m] = \sum_{x=0}^{N-1} \mathbf{p}[x]\,\mathbf{q}[m+x] \quad m = 0, 1 \ldots N-1$$

As we now have 2-D images, their correlation is

$$\mathbf{CORR} = \mathbf{P} \otimes \mathbf{Q} = \sum_{x=0}^{C-1}\sum_{y=0}^{R-1} \mathbf{P}[x,y]\,\mathbf{Q}[m+x,n+y] \tag{4.26}$$
$$m = 0, 1 \ldots C-1; n = 0, 1 \ldots R-1$$

which is the same as convolution without the axis inversion. Convolution is actually correlation if the image or the template is symmetric, because the inversion process has no effect. As this is compact notation, an alternative form for a template of the same dimensions as the image, $C \times R$, the direct implementation of image correlation is

$$CORR_{x,y} = \sum_{x=0}^{C-1}\sum_{y=0}^{R-1} P_{x,y} \times Q_{\mathrm{mod}(x+x',C-1),\mathrm{mod}(y+y',R-1)} \tag{4.27}$$
$$x' \in 0, 1 \ldots C-1, R; y' \in 0, 1 \ldots R-1$$

As in Equation 3.48 we can invoke the convolution theorem and use the FFT to speed image correlation as

$$\mathbf{P} \otimes \mathbf{Q} = \mathcal{F}^{-1}\left(\mathcal{F}(\mathbf{P}) . \times \overline{\mathcal{F}(\mathbf{Q})}\right) \tag{4.28}$$

where $\overline{\mathcal{F}(Q)}$ denotes the complex conjugate, or we can implement this by inverting the transform around both axes as

$$P \otimes Q = \mathcal{F}^{-1}(\mathcal{F}(P) . \times \mathcal{F}((C-1, R-1) - Q)) \qquad (4.29)$$

where the term $(C-1, R-1) - Q$ denotes inversion of the image Q along both axes

$$\mathcal{F}((C-1, R-1) - Q) = FQ_{C-1-i, R-1-j} \quad i = 0, 1 \ldots C-1; j = 0, 1 \ldots R-1$$
$$(4.30)$$

4.4.3.2 Template correlation/ template matching

As in convolution, there is also *template correlation* for templates, which are smaller than the image and this is the most usual use of image correlation. There are two main approaches that deploy image/template correlation:

1. when the template is symmetric and the result of convolution is the same as that of correlation because there is no inversion; and
2. *template matching*, because matching is the function of correlation, can be used to locate the position of the template in an image (and thus to find the object contained within the template).

The direct implementation of direct template correlation/matching is given as

$$T\_CORR_{x', y'} = \sum_{x=0}^{C_T-1} \sum_{y=0}^{R_T-1} P_{x'-\frac{C_T}{2}+x, y'-\frac{R_T}{2}+y} \times w_{x,y} \quad \begin{bmatrix} x' \in C_T/2, C-1-C_T/2 \\ y' \in R_T/2, R-1-R_T/2 \end{bmatrix}$$
$$(4.31)$$

It is called matching because if the template values w are set to 1.0 for the points in the target shape and zero otherwise, and the image is thresholded so as to make the significant points 1.0 (and the other points are set to values less than that – preferably to zero, though that is not always possible), then the result of correlation at a point is a sum of the points in the image that matched those in the template placed at that point. If either the image points are zero or the template values are zero, then their product will not contribute to the correlation summation. As ever this is programming artifice and a different way to achieve template matching is

$$T\_CORR_{x', y'} = \sum_{x=0}^{C_T-1} \sum_{y=0}^{R_T-1} \left\{ \begin{array}{l} 1 \ P_{x'-\frac{C_T}{2}+x, y'-\frac{R_T}{2}+y} = w_{x,y} \\ 0 \ \text{otherwise} \end{array} \right. \quad \begin{bmatrix} x' \in C_T/2, C-1-C_T/2 \\ y' \in R_T/2, R-1-R_T/2 \end{bmatrix}$$

Again, template correlation can be implemented using Fourier when the templates are of sufficient size. In fact, a small template of size 3×3 or 5×5, apart from template convolution is not much use for anything except for finding a full stop or a minus sign so by the analysis in Section 4.4.2, a Fourier implementation of image correlation is often used for template matching.

Template matching has well-known performance attributes in that it can tolerate very well indeed the effects of noise and occlusion (when sections are hidden). When the template is an exact match to the target shape, it is a form of *matched template filtering*. Matched filtering is well known for its optimal properties – a binary line receiver can use an integrate-and-dump receiver, which is a matched filter for binary signals. Because the matching process is a counting one, then if either the noise or the occlusion reduces the count below some threshold, then the shape is not detected.

4.4.3.3　Finding objects by template correlation/matching

The process of finding objects by template correlation/matching is shown in Figure 4.22 where it is applied to find the camels originally shown in Figure 4.15. Because the direct implementation is used (Equation 4.27) only small images are used here. A crop of Figure 4.15 contains four (invariably grumpy) camels in Figure 4.22(a), which is an image of resolution 163×269. We shall detect a different camel, Figure 4.22(b), which is cropped from the same original image and is a 17×17 template. The images are then thresholded to give Figures 4.22(c) and (d) where a slightly different threshold (finding the darker points) was used in each, primarily for illustrative purposes (sorry camels, we lost most of your legs – there are optimal thresholding techniques, which aim to find the threshold automatically [Nixon & Aguado, 2019]). We then correlate/match the template with the image to give the positions of the best match of the template, via template matching in Figure 4.22(e) and image matching via Fourier in Figure 4.22(f) (the difference between the two results is the borders of the image, the same as with image convolution in Figure 4.21(d)). Clearly, we have detected the four camels as there are four clear peaks in each result image – each peak is determined by the points that match between the thresholded template and the thresholded image. The peak is less pronounced for the camel from the left of centre, because that has fewer points in the thresholded image Figure 4.22(c). The implementation of template correlation ignores the image borders, whereas image correlation (Equation 4.29) assumes replication (deploying Equation 4.28 also gives the same result as in Figure 4.22(f)). Given the size of the template, the Fourier implementation is preferred for reasons of speed. A more reliable approach could use edge detection in preference to thresholding, and that would lead to a similar result and without the need to choose a threshold, so for completeness

FIGURE 4.22 Applying correlation/matching: (a) camels; (b) a different camel; (c) thresholded images of camels; (d) thresholded image of a different camel; (e) template correlation/matching; (f) image correlation/matching; (g) edge image of camels; (h) edges of template

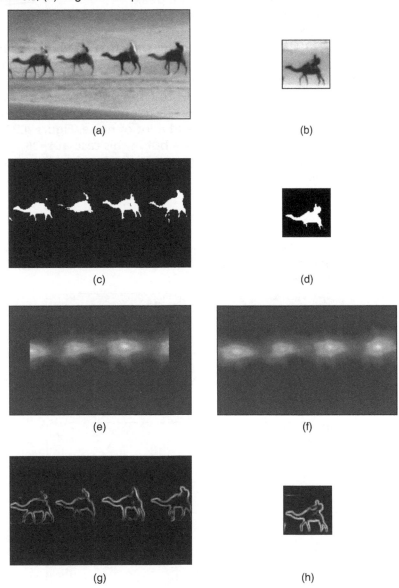

(a)

(b)

(c)

(d)

(e)

(f)

(g)

(h)

(and because Professors of Computer Vision would avoid thresholding here) the edge images are shown in Figures 4.22(g) and (h). A template that can be convolved with an image to detect edges is described later in Section 6.3.4.

The matching process is actually very resilient to noise, as there is much averaging in the process. The position of the best match is revealed by the location of the peak in the result of correlation: for Figure 4.22(f)

> So we can find camels in a sandstorm. Very useful!

the peak is located at coordinates x=168, y=64. With a slightly noisy image, Figure 4.23(a), the peak is found at the same location (though the peak is less clear due to the noisy points) and is shown with a cross in Figure 4.23(d). With a much more noisy image, Figure 4.23(b), the peak is found at x=168, y=63 (reliably, within multiple instantiations of random noise) even when the resulting correlation is more evenly distributed in Figure 4.23(e). When we add a lot of noise, Figure 4.23(c), the peak is not found at the same place but in this case at x=35, y=70 (where a different camel is found, Figure 4.23(f)). Fourier allows us to find shapes faster.

FIGURE 4.23 Correlation/matching in noisy images: (a) slight noise; (b) more noise; (c) very noisy; (d) correlation result for (a); (e) correlation result for (b); (f) correlation result for (c)

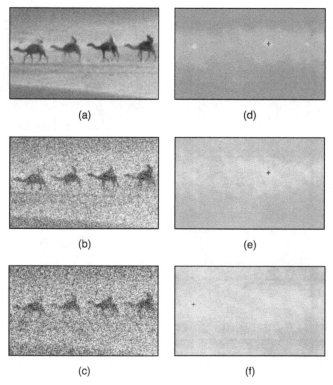

(a)   (d)

(b)   (e)

(c)   (f)

4.4.4 Filtering

4.4.4.1 Low- and high-pass filtering

Sections 2.6 and 3.5 considered the filtering of 1-D signals using low-pass and high-pass filters. We shall do the same now for images, starting with low-pass filtering. In order to display the transforms and filters, we shall centre the transforms for display. We then achieve low-pass filtering by retaining components within a circle of radius equal to the cut-off frequency and which is centred on the d.c. component.

$$
\mathbf{LP} = low\_pass\,(\mathbf{FP}, cut\_off) = \begin{cases} FP_{u,v} & \left(u - \dfrac{C}{2}\right)^2 + \left(v - \dfrac{R}{2}\right)^2 < cut\_off^2 \\ 0 & \text{otherwise} \end{cases}
$$

$$(4.32)$$

Conversely, high-pass filtering obliterates the low frequency components selecting those which were zapped in low-pass filtering, as

$$
\mathbf{HP} = high\_pass\,(\mathbf{FP}, cut\_off) = \begin{cases} FP_{u,v} & \left(u - \dfrac{C}{2}\right)^2 + \left(v - \dfrac{R}{2}\right)^2 \geq cut\_off^2 \\ 0 & \text{otherwise} \end{cases}
$$

$$(4.33)$$

These filters and their effects are shown in the images reconstructed from the filtered transforms, for example, $\mathcal{F}^{-1}\,(\mathbf{LP})$, in Figure 4.24. The desired frequencies are retained (broadly, the low-pass filter retains structure and the high-pass filter retains change), but there are effects of the sampling in the frequency domain. As in the 1-D counterpart, Section 3.5, the filters are equivalent to a pulse in the frequency domain and this leads to the curious artefacts – the *ringing* – in the filtered images, which can be seen clearly in the low-pass filtered image and in the high-pass filtered image. The operation is simple, but that is its main merit.

In order to improve matters, we previously used a Gaussian filter because the transform of a Gaussian is also a Gaussian so the function is smooth in the spatial and frequency domains. Given a centred Fourier transform we can then achieve low-pass Gaussian filtering as

$$
\mathbf{LPG} = low\_pass\_Gauss\,(\mathbf{FP}; \sigma) = FP_{u,v}\,e^{\dfrac{-\left(\left(u-\frac{C}{2}\right)^2 + \left(v-\frac{R}{2}\right)^2\right)}{2\sigma^2}}
$$

$$(4.34)$$

FIGURE 4.24 Image filtering: (a) original image; (b) low-pass filter components; (c) high-pass filter components; (d) low-pass filtered image; (e) high-pass filtered image

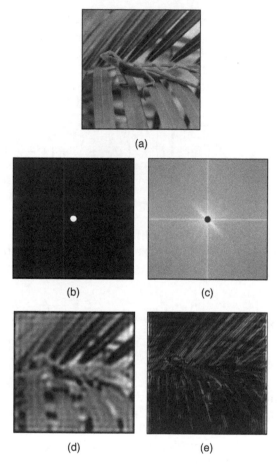

where the centre point is at coordinates $\left(\frac{C}{2}, \frac{R}{2}\right)$. Conversely, high-pass filtering obliterates the low-frequency components,

$$\textbf{HPG} = high\_pass\_Gauss\,(\textbf{FP}; \sigma) = FP_{u,v}\left(1 - e^{\dfrac{-\left(\left(u-\frac{C}{2}\right)^2 + \left(v-\frac{R}{2}\right)^2\right)}{2\sigma^2}}\right) \quad (4.35)$$

When these are used to filter images, shown in Figure 4.25, the arte-facts disappear and the low-pass image is smoothed, as required. The effects of ringing have disappeared because the Fourier transform of

FIGURE 4.25 Gaussian image filtering: (a) Gaussian low-pass filter;
(b) Gaussian high-pass filter

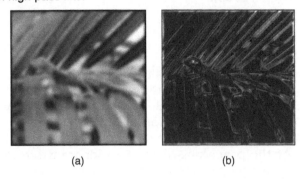

(a)   (b)

the Gaussian function is another (smooth) Gaussian function with different variance (as in a 2-D version of Equation 3.22). This imposes little extra computation because the entire process involves floating-point computation. By the results, it is clearly well worth adding.

4.4.4.2 Unsharp masking

There is a standard procedure aimed to improve the quality of images which is based on filtering. This can be achieved optically when processing photographs, but is now more likely to be achieved digitally. The process is *unsharp masking*, which by its name aims to obscure blurred image data. One way to achieve this is to subtract a low-pass filtered version of an image

$unsharp\_masked\,(\mathbf{P}; \sigma, um)$

$$= \mathbf{P} \times (1 + um) - \mathcal{F}^{-1}\left(low\_pass\_Gauss\,(\mathbf{FP}; \sigma)\right) \times um \qquad (4.36)$$

where the amount of filtering is controlled by σ and the amount of subtraction is controlled by the factor um. This can require image normalisation afterwards. An equivalent process is to reinforce high frequency information

$unsharp\_masked\,(\mathbf{P}; \sigma, um)$

$$= \mathbf{P} \times (1 - um) + \mathcal{F}^{-1}\left(high\_pass\_Gauss\,(\mathbf{FP}; \sigma)\right) \times um \qquad (4.37)$$

Some implementations include thresholding to determine significance of components. The process can be achieved by using template convolution, and a template can be derived based on Figure 4.18(c) or templates equivalent to high-pass filtering which will be covered more when differentiation (which is known as edge detection in image processing/computer vision) is considered within applications of the Fourier transform in Chapter 6. There is still relatively recent interest in unsharp marking, more for doyens of image processing, e.g. [Deng, 2010; Polesel et al., 2000].

4.5 SUMMARY

We have now covered the main theory of the Fourier Transform. Chapter 2 started with the 1-D continuous version, before the discrete version (and the FFT) in Chapter 3. In this chapter we have used the continuous and discrete versions for 2-D images, with theory extending beyond Chapters 2 and 3 when it needed extension to two dimensions. Some of the properties of the Fourier transform are actually much clearer in images/pictures largely since we can see the results (we can only hear or plot the results for one-dimensional analysis) and consider them at leisure and in detail. OK, we had to know some of the things that go wrong with our own sight and nothing is perfect. Seeing the results helps particularly with shift invariance and to showing the importance of phase, as well as to filtering and getting rid of the noise. Fourier used sine and cosine functions since, well, he was the first but also because they are intrinsically smooth and yield to calculation. There are however many other functions possible, and these lead to important advantages in different areas such as image coding from which pretty much everyone in the world has benefitted from. Let's go and have a look.

CHAPTER 5

Variants of the Fourier Transform

Fourier was the first to notice that we can represent signals with sine and cosine waves as the basis functions. This was a fantastic notion and the sine and cosine basis functions are smooth and yield to analysis. There is of course an infinite selection of basis functions, and they provide differing insights as to the composition of a signal. Some of them are nice ideas but utterly useless, and some have particular advantages. We shall start with the cosine transform since it is the one that pretty much everyone has used, largely because it is commonly used to compress signals and images to use less storage. Let us emphasise the point about generality in key point 8.

Key point 8

The Fourier transform is based on sine and cosine waves and uses only one of an infinite number of possible basis function sets.

As we noted when studying low-pass filtering in Section 4.4, a lot of the image information is contained in the low frequencies. If we can intelligently select the important frequencies, rather than just the low ones, then the image can be stored using much less space. We shall find that Fourier is too general for this, and it is better to change the basis functions. The general approach in this Chapter will be to start with the continuous version of a transform and cover any general points of interest. We shall then move to the 1-D discrete transform, before the 2-D one which is where we can see some of the basic tenets of the approach, using images.

5.1 COSINE AND SINE TRANSFORMS, INCLUDING THE DISCRETE COSINE TRANSFORM

5.1.1 1-D continuous transforms

The 1-D *continuous cosine transform* does not use complex numbers and is defined as

$$Cp\,(f) = \int_{-\infty}^{\infty} p\,(t)\cos\,(2\pi ft)\,dt \qquad (5.1)$$

By virtue of the cosine function it is an even function, so $Cp\,(f) = Cp\,(-f)$. There is also a *sine transform*

$$Sp\,(f) = \int_{-\infty}^{\infty} p\,(t)\sin\,(2\pi ft)\,dt \qquad (5.2)$$

but it leads nowhere except to the reconstruction

$$p\,(t) = \int_{0}^{\infty} Cp\,(f)\cos\,(2\pi ft)\,df + \int_{0}^{\infty} Sp\,(f)\sin\,(2\pi ft)\,df \qquad (5.3)$$

These will not be implemented here as they are largely of academic value only. They do lead to the discrete version and that is the one you and I use, probably every day (or even every hour!).

5.1.2 1-D discrete cosine and sine transforms

5.1.2.1 Discrete cosine transform and compression

The *Discrete Cosine Transform* (*DCT*) [Ahmed et al., 1974] is a real transform that has great advantages in compression, motivating its use in coding. The most common version of the 1-D DCT calculates transform components **Cp** from a signal **p** as

$$Cp_u = \sum_{x=0}^{N-1} p_x \cos\left(\frac{\pi}{2N}(2x+1)u\right) \quad u = 0, 1 \ldots N-1 \qquad (5.4)$$

and its implementation is given in Code 5.1, noting as usual that since arrays in Matlab are indexed from 1 to N rather than 0 to $N-1$ then in the cos function u becomes u-1, and $2x+1$ becomes 2*x-1.

Code 5.1 Discrete Cosine transform

```
function trans = DCT(vector)

points=length(vector); %get dimensions
trans(1:points)=0; %initialise output transform
```

```
%compute transform
for u = 1:points %address all components
    sum=0;
    for x = 1:points %address all points
        sum=sum+(vector(x)*cos((pi/(2*points))*
                                (2*x-1)*(u-1))); %Eq 5.4
    end
    trans(u)=sum;
end
```

The other versions of the DCT largely differ in the scaling coefficients (of which there are none here) and the values for the start and end coefficients (which can affect orthogonality). The version here is known as *DCT-II*. When this is applied to the data in Figure 5.1(a) we derive the coefficients in Figure 5.1(b). Clearly, the transform does not involve complex numbers so it results in real values; most of the DCT transform coefficients are near zero (and can be ignored or discarded, hence the possibility of compression).

By its use of the cosine function, and its formulation, the repetition for the DCT is not the same as for the discrete Fourier transform (DFT). Since $\cos(-a) = \cos(a)$ the transform assumes that the data, Figure 5.2(a), repeats as a mirror image shown in Figure 5.2(b). The data to which the DCT is assumed to be applied is shown in Figure 5.2(c). This avoids the discontinuities that arise from replication in the DFT (Figure 3.15), since the end values are the same (at integer multiples of *N*). Also, the lack of symmetry in the DCT implies that there are twice as many spectral estimates as in the DFT (it is periodic for *2N*), though many of these can be very small.

> These assumptions are implicit in the formulation and are only readily apparent when it is applied, unless you're a maths junkie!

FIGURE 5.1 Applying the discrete cosine transform: (a) signal **p**; (b) DCT **Cp**

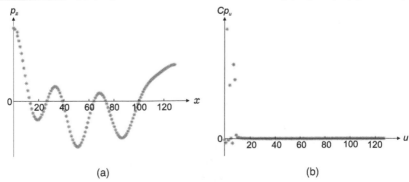

(a)   (b)

FIGURE 5.2 Replication in the discrete cosine transform: (a) original signal, $p_x|x \in 0, 127$; (b) extension implicit in DCT, $p_{2N-x-1} = p_x$ $N = 128$; (c) replication in DCT

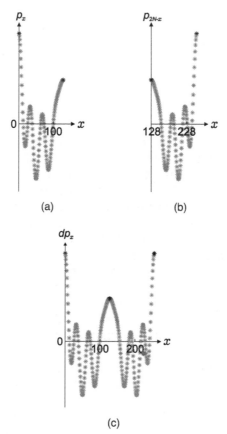

(a)   (b)

(c)

No transform is of practical use without an inverse. The *inverse DCT* is

$$p_x = \frac{1}{N} \sum_{u=0}^{N-1} Cp_u w_u \cos\left(\frac{\pi}{2N}(2u+1)x\right) \quad x = 0, 1 \ldots N-1 \quad (5.5)$$

where the weighting factor affects the zero-order coefficient

$$w_u = \begin{cases} 1 & u = 0 \\ 2 & \text{otherwise} \end{cases}$$

When this inverse DCT is used to invert the transform in Figure 5.1(b), the maximum difference between an original and a reconstructed component is $O(10^{-14})$, which is arithmetic error. So let us consider the compression in more detail. We just noted that most of the DCT transform coefficients in Figure 5.1(b) are near zero.

So we can reconstruct the signal from the most significant transform components by thresholding the transform at different values (retaining those components whose magnitude exceeds a chosen threshold). We shall reconstruct the signal in Figure 5.1(a) from some of the transform components of Figure 5.1(b) using thresholds of 20, 10 and 1. With these thresholds 2, 4 and 7 components, respectively, are retained which represent only 2%, 3% and 5% of the transform data, respectively. We can see that even with this enormous compression, we can reconstruct the signal very well, with low error and the differences between Figures 5.3(b) and (c) are very slight and their errors (Figures 5.3(e) and (f), respectively) are very small. Rather than threshold the components we can also see how the components themselves affect the reconstruction accuracy, shown in Figure 5.3(g). This shows the Root Mean Square (RMS) error

$$\left(\sqrt{\sum_{x \in 0, N-1} \left(\text{reconstructed}\, p_x - \text{original}\, p_x \right)^2 / N} \right)$$ that accrues when

increasingly more frequency components are included in the reconstruction via the inverse DCT. The first error point is the error when only the first frequency component is included in the reconstruction which is when the error is largest; the second error is when the first and second components are included. The reconstruction error reduces to near zero when the first 10 of the 128 frequency components are included, confirming that few components are needed to reconstruct the signal accurately, those components shown in Figure 5.1(b). The error does not reduce when there is no component at that frequency.

The signal we have been using so far is rather clean, so let us look at the Hard Day's Night Chord, shown in Figure 5.4(a), with DCT transform shown in Figure 5.4(b) together with the regions of smaller components that will be omitted (excluded) by thresholding. When we reconstruct from the largest components with absolute magnitude greater than 20, then we retain only 21 of 2048 components, but the mean absolute reconstruction error of around 20% confirms that the signal in Figure 5.4(c) is a poor reconstruction of Figure 5.4(a). When the 93 components of magnitude exceeding 10 are retained, the reconstruction is better (Figure 5.4(d)) with mean absolute error (MAE) of 11%. Reconstruction is best when 588 components exceeding 1 are retained in Figure 5.4(e), with a 2% MAE and the reconstructed signal is visually similar to the original version in Figure 5.4(a).

5.1.2.2   Basic coding

We will discuss *coding* in detail in the next chapter, as it attracts many varieties of frequency domain analysis. Essentially, coding aims to save storage by eliminating data. This was noted in Figures 5.3(b) and (c)

FIGURE 5.3 Reconstruction via the inverse DCT: (a) reconstruction by retaining components ≥ 20; (b) reconstruction by retaining components ≥ 10; (c) reconstruction by retaining components ≥ 1; (d) error in (a); (e) error in (b); (f) error in (c); (g) RMS error when reconstructing using increasing numbers of components

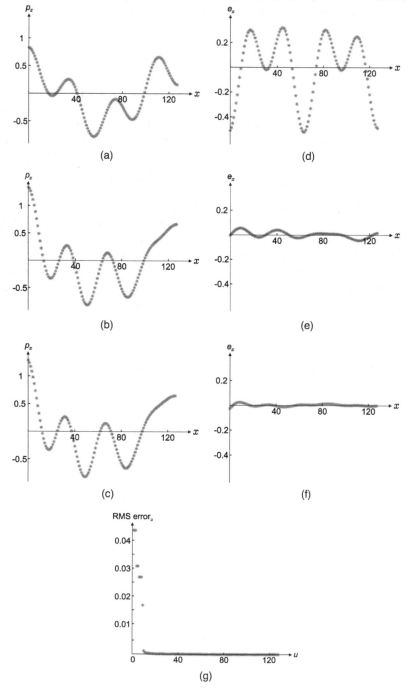

FIGURE 5.4 Reconstruction of a real signal from DCT components: (a) signal; (b) transform by DCT; (c) reconstruction with components ≥ 20; (d) reconstruction with components ≥ 10; (e) reconstruction with components ≥ 1

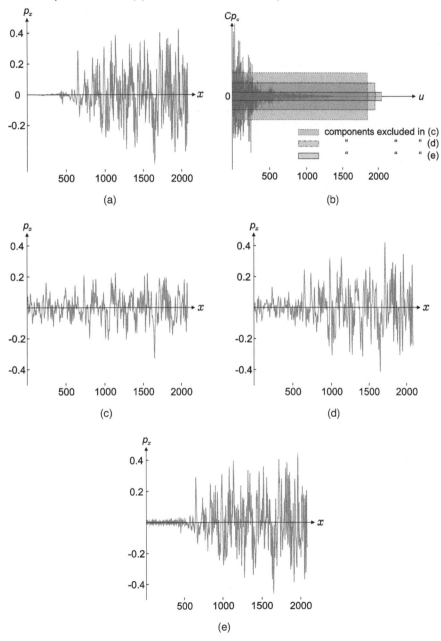

where one used nearly twice as many bits as the other (OK, there weren't many components in the first place, but we are discussing principles here). Though Figure 5.3(c) contained more components, it looked very similar to the original signal, as confirmed by the error analysis, Figure 5.3(f). Clearly, all components are needed (*lossless compression*) if the representation is to be exact; we can code a signal with *lossy compression* when we just want to hear it; and the loss does not matter if it concerns material that is imperceptible to a user or an application. One way of viewing coding is that it is a form of intelligent filtering: we select the components necessary to represent a signal. One explanation is that smooth areas need little coding since nothing changes; this can also phrased by considering that coding removes correlations. Note that any *artefacts* (unsmooth bits) are quite likely to be noticed in smooth areas and affect the perceived quality of the coding. Another explanation with regards to transform-based coding is that large isolated components are very likely to be retained, but where components cluster they can be compressed since there is little perceptual difference between similar frequencies. Naturally, coding is more cannier than this, as we shall find later.

The compression/coding properties of the DCT arise from the cosine basis functions, whereas Fourier is too general. To test this, we shall repeat the analysis of Figure 5.4, using the Fourier transform as well. The results tabulated in Table 5.1 show that though the error is largely similar for both transforms at different thresholds, the number of components retained is considerably less for the DCT and the compression is much greater, storing fewer components for similar performance (and without complex numbers!). Another explanation for this is that the DCT offers the closest approximation to *principal component analysis* [Weilong et al., 2005], which in machine learning is well known for its ability to compress data. We shall look at coding again when we compare some different image transforms in Section 5.4. There we shall use a fixed number of components and consider the difference in error.

> One can suggest that a converse of coding is possible: that filtering is dumb compression – but that's a bit daft.

5.1.2.3 Relationship between the DCT and the DFT

It is now propitious to investigate the relationship between the DFT and the DCT, since it gives a way to calculate the DCT via the FFT. To analyse this, we first double the length of the data, by reflecting around p_{N-1} (previously shown in Figure 5.2) as

$$dp_x = \begin{cases} p_x & x = 0, 1 \ldots N - 1 \\ p_{2N-x-1} & x = N, \ldots 2N - 1 \end{cases} \tag{5.6}$$

TABLE 5.1 Compressing the signal of Figure 5.4(a) using the DFT and the DCT.

Transform	Threshold	Number of components	Compression/ compaction	Mean absolute error	Maximum error
	1	588	72%	2%	9%
DCT	5	198	90%	6%	31%
	10	93	96%	11%	52%
	1	805	61%	1%	11%
DFT	5	298	86%	5%	30%
	10	176	92%	9%	46%

The DFT of the double-length data is

$$Fdp_u = \frac{1}{2N} \sum_{x=0}^{2N-1} dp_x e^{-j\frac{2\pi}{2N}xu} \quad u = 0, 1 \ldots 2N - 1$$

which is in two halves

from dp_x, Equation 5.6

$$Fdp_u = \begin{cases} Fp_u & u = 0, 1 \ldots N - 1 \\ Fp_{N-u} & u = N, \ldots 2N - 1 \end{cases}$$

For the two halves

$$Fdp_u = \frac{1}{2N} \sum_{x=0}^{N-1} p_x e^{-j\frac{2\pi}{2N}xu} + \left(\frac{1}{2N} \sum_{x=N}^{2N-1} p_{2N-x-1} e^{-j\frac{2\pi}{2N}xu} \right)$$

and by symmetry (or by sleight of hand?)

$$Fdp_u = \frac{1}{2N} \sum_{x=0}^{N-1} p_x e^{-j\frac{2\pi}{2N}xu} + \left(\frac{1}{2N} \sum_{x=0}^{N-1} p_x e^{-j\frac{2\pi}{2N}(2N-x-1)u} \right)$$

Since $e^{-j2\pi u} = 1$

$$Fdp_u = \frac{1}{2N} \sum_{x=0}^{N-1} p_x e^{-j\frac{\pi}{N}xu} + \frac{1}{2N} \sum_{x=0}^{N-1} p_x e^{j\frac{\pi}{N}xu} e^{j\frac{2\pi}{2N}u}$$

Factorising $e^{j\frac{2\pi}{2N}u} = e^{j\left(\frac{\pi}{2N}+\frac{\pi}{2N}\right)u}$ gives

$$Fdp_u = e^{j\frac{\pi}{2N}u} \left(\frac{1}{2N} \sum_{x=0}^{N-1} p_x \left(e^{-j\frac{\pi}{N}xu} e^{-j\frac{\pi}{2N}u} + e^{j\frac{\pi}{N}xu} e^{j\frac{\pi}{2N}u} \right) \right)$$

Simplifying, $Fdp_u = e^{j\frac{\pi}{2N}u} \left(\frac{1}{2N} \sum_{x=0}^{N-1} p_x \left(e^{-j\frac{\pi}{2N}(2x+1)u} + e^{j\frac{\pi}{2N}(2x+1)u} \right) \right)$

thus $\quad Fdp_u = e^{j\frac{\pi}{2N}u} \frac{1}{N} \sum_{x=0}^{N-1} p_x \cos\left(\frac{\pi}{2N}(2x+1)u \right)$

By Equation 5.4, the DFT can be computed as

$$Fdp_u = \frac{1}{N}e^{j\frac{\pi}{2N}u}Cp_u \tag{5.7}$$

and the DCT can be computed from the DFT (and the FFT) as

$$Cp_u = Ne^{-j\frac{\pi}{2N}u}Fdp_u \tag{5.8}$$

There appears to be other versions of this relationship (largely varying in the scaling coefficient, as ever with the DFT; the formulation here is consistent with the definition of the DFT in Equation 3.6) so the implementation of Equation 5.8 is given in Code 5.2. This fits with the definition of the DCT given in Equation 5.4. Here the error in the first three elements of difference between the components computed by the DCT (Cp) and those by Equation 5.8 (nCp) appears to have imaginary components whereas it should be purely real, but it is actually near zero (10^{-14}). The average error shows this more clearly as the error in the real part differs from that in the imaginary part, and both are because of the computer arithmetic. This is reproducibility: Equation 5.8 clearly works (even if its derivation is a bit dodgy!).

Code 5.2 Computing the DCT from the DFT (and FFT)

```
N=128; %use p
for i=1:N
    p(i)=0.5*cos((2*pi/N)*i)+0.25*cos((2*pi/N)*2*i)+...
        0.35*cos((2*pi/N)*3.5*i)+0.25*cos((2*pi/N)*4*i);
end
plot(p,'*')
dp=p; %Eq 5.6 top
dp(N+1:2*N)=p(N:-1:1); %Eq 5.6 bottom
Fp=DFT(dp); %if you use fft need to scale by 1/length(dp)
Cp=DCT(p); %take DCT
for u=1:length(dp)
    nCp(u)=N*exp(-j*pi*(u-1)/(2*N))*Fp(u); %Eq 5.8
end
nCp(1:3)-Cp(1:3) %difference for first three elements
ans = 1.0e-14 * -0.1998 + 0.0000i 0.6883 + 0.7133i
        0.0000 + 0.5107i
aver=0; %let's work out the average difference
for u=1:length(p)
    aver=aver+(nCp(u)-Cp(u));
end
aver=aver/length(p) %average difference
aver = -1.1824e-14 + 5.9417e-15i
```

There are two implications to Equation 5.8:

1. the relationship between the DCT and the DFT is for a sequence of double the length; and
2. we can compute the DCT via the FFT.

The latter of these is more for convenience, though it is much faster than the basic DCT. If this was a book on the DCT we would continue with a fast DCT algorithm (which would save N multiplications over when using the FFT). But it ain't, and we won't.

5.1.2.4 Other properties of the DCT

Beyond compression and being real numbers, there are few advantages of the DCT. There is no shift invariance (or, alternatively, time shift is not lost), but it does obey superposition and is thus a linear transformation. Other properties are intrinsic to frequency domain analysis so these properties are shared with the DFT, like rotation and scaling. Symmetric convolution is possible with the DCT [Martucci, 1994]. As just mentioned, there are also fast DCT algorithms (e.g. for error analysis see [Tun & Lee, 1993]) which can be implemented in $O\left(N log_2 N\right)$). These are covered nicely in books focussed on the DCT (e.g. [Britanak et al., 2010]).

5.1.2.5 Discrete sine transform

There is also a *Discrete Sine Transform* (DST),

$$Sp_u = \sum_{x=0}^{N-1} p_x \sin\left(\frac{\pi}{2N}(2x+1)u\right) \quad u = 0, 1 \ldots N-1 \qquad (5.9)$$

This is equivalent to the magnitudes of the imaginary parts of the DFT (the DCT is the real parts) of a sequence of twice the length and has odd symmetry via the sine function, unlike the DCT, which has even symmetry.

The *inverse DST* is

$$p_x = \sum_{u=0}^{N-1} Sp_u w_u \sin\left(\frac{\pi}{2N}(2u+1)x\right) \quad u = 0, 1 \ldots N-1 \qquad (5.10)$$

where the weighting factor affects the zero-order coefficient

$$w_u = \begin{cases} 1 & u = 0 \\ 2 & \text{otherwise} \end{cases}$$

Note that boundary conditions can affect the applications of the DST.

5.1.3 2-D discrete cosine transform

The *2-D DCT* definition for computing spectral components **CP** from an image **P** is:

$$CP_{u,v} = \sum_{x=0}^{N-1}\sum_{y=0}^{M-1} P_{x,y}\cos\left(\frac{\pi}{2N}(2x+1)u\right)\cos\left(\frac{\pi}{2M}(2y+1)v\right) \qquad (5.11)$$

This is the *2-D DCT II* and again there are variants of the definition of the DCT and we are concerned only with principles here (the differences are largely in the scaling factors). For efficiency, the transform is computed using separability, first by transforming the rows and then the columns, as

$$CP_{u,v} = \sum_{x=0}^{N-1}\left(\left(\sum_{y=0}^{M-1} P_{x,y}\cos\left(\frac{\pi}{2M}(2y+1)v\right)\right)\cos\left(\frac{\pi}{2N}(2x+1)u\right)\right)$$

If you do not use a fast version, you have to wait a long time. The DCT of an image in comparison with its DFT is shown in Figure 5.5. Clearly, the DCT components are focussed on the top left corner, noting that there are four times as many components of the DCT than there are for the DFT (the relationship noted in Section 5.1.2), though most of these are near zero. The DFT image has not been derived by centring, to emphasise this.

The *inverse 2-D DCT (2-D DCT IV)* is defined by

$$P_{x,y} = \sum_{u=0}^{N-1}\sum_{v=0}^{M-1} CP_{u,v}w_u w_v\cos\left(\frac{\pi}{2N}(2x+1)u\right)\cos\left(\frac{\pi}{2M}(2y+1)v\right) \qquad (5.12)$$

where

$$w_u, w_v = \begin{cases} 1 & \text{if } u = 0 \text{ or } v = 0 \\ 2 & \text{otherwise} \end{cases}$$

and this is again computed using separability.

The action of the 2-D DCT can be visualised using its basis functions. There are 64 basis functions in 8×8 blocks (not including the single-pixel border around each block) shown in Figure 5.6. Here the first component in the top left has no change (d.c.) and the frequencies increase in horizontal and vertical senses until the maximum frequency components are reached at the bottom right. To compute the transform, these basis functions can be pre-computed and then convolved with the image. As we shall find in the applications, much use is made of the excellent coding properties of the DCT, and more detailed descriptions can be found [e.g. Rao & Yip, 2014].

There is also a *2-D DST*, but it appears to be little used.

FIGURE 5.5 Comparing the transform by the DCT with that by the DFT: (a) original image; (b) |2-D DFT|; (c) 2-D DCT

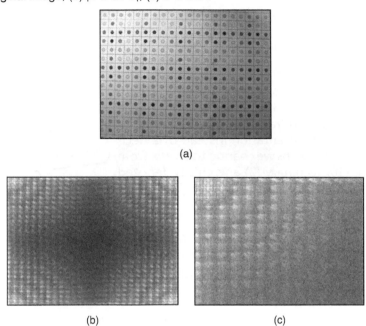

(a)

(b)   (c)

FIGURE 5.6 DCT basis functions

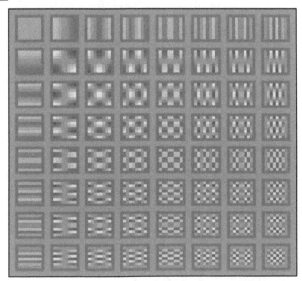

5.2　WALSH–HADAMARD TRANSFORM

5.2.1　Walsh transform

5.2.1.1　The 1-D transform

It is nearly a century since Walsh first proposed a binary set of basis functions [Walsh,
1923]. It is binary and so we shall start with the
1-D *discrete Walsh transform*. We first need to
use a binary representation instead of a decimal one. That means we change the *radix* from
decimal (10) to binary (2) and this is denoted
using suffices for a number p as

> Calling Walsh 'discrete' is largely superfluous since binary signals do not exist as continuous versions, except in maths – even a Bugatti can't make it to a ton in zero time

$$(p)_{10} = (p)_2$$

(Using 10 as a suffix for decimal can be a bit confusing since it also
represents 2 as a binary number. Writing it as $(p)_{10_{10}} = (p)_{2_{10}}$ would be
unnecessarily pedantic. Doh, we just wrote it!) The binary number is a
string of bits b_k

$$(p)_{10} = \left(b_{N-1}(p) \dots b_1(p) b_0(p)\right)_2$$

If p is in the range $0 \dots 255$ it is represented by an 8-bit number where
the position of the nth bit (from the least significant bit [LSB], which
is the rightmost bit) represents 2^{n-1}, for example,

$$(134)_{10} = (10000110)_2$$

which is $134_{10} = 128 + 4 + 2$. For more details on binary numbers, con-
sult the opuscule (Ha! That means small book) [Nixon, 2015]. Sampled
signals have a bit length that is fixed by an analogue to digital converter (or by a computer). The 1-D discrete Walsh transform **Wp** of a
sampled signal **p** is then

$$Wp_u = \frac{1}{N} \sum_{x=0}^{N-1} p_x \times \prod_{k=0}^{Nb-1} (-1)^{b_k(x)b_{Nb-1-k}(u)} \quad u = 0, 1 \dots N - 1 \quad (5.13)$$

where $Nb = \log_2 N$ and the basis functions $\prod_{k=0}^{Nb-1} (-1)^{b_k(x)b_{Nb-1-k}(u)}$ are
either $+1$ or -1. These are a set of binary orthonormal basis functions,
in a manner similar to the Fourier transform basis. The product term
can be simplified as

$$Wp_u = \frac{1}{N} \sum_{x=0}^{N-1} p_x \times (-1)^{\sum_{k=0}^{Nb-1} b_k(x)b_{Nb-1-k}(u)} \quad (5.14)$$

Denoting the signal and its transform as column vectors

$$\mathbf{p} = \begin{bmatrix} p_0 \\ p_1 \\ \vdots \\ p_{N-1} \end{bmatrix} \quad \text{and} \quad \mathbf{Wp} = \begin{bmatrix} Wp_0 \\ Wp_1 \\ \vdots \\ Wp_{N-1} \end{bmatrix}$$

then the transform can be written using a Walsh transform matrix **W** as

$$\mathbf{Wp} = \mathbf{W} \times \mathbf{p} \qquad (5.15)$$

where the elements of **W** are given by

$$W_{u,x} = (-1)^{\sum_{k=0}^{Nb-1} b_k(x) b_{Nb-1-k}(u)}$$

The functions with increasing frequency have a greater number of zero crossings, where the signal changes from $+1$ to -1. This is known as *sequency* and the first eight functions are shown in Figure 5.7 where the frequency/sequency increases by a factor of two between adjacent functions, and the signals with high sequency are later in the order.

A signal and its Walsh transform components are shown in Figure 5.8. As ever, the transform coefficients peter out, affording the possibility of compression and thus of coding. By visual comparison with the DCT in Figure 5.1(b) the Walsh transform in this case shows less compression/compaction, and this is generally found to be the case.

The transform basis, the Walsh basis functions **W** for an 8-bit system are shown in Figure 5.9. (This is an 8-bit system since the code complexity in Matlab is high for smaller bit lengths.) Figure 5.9 shows the binary functions increasing in frequency from d.c. (top left) to the highest frequency (bottom right). The high-frequency/sequency functions predominate in the lower right of Figure 5.9. It is perhaps non-intuitive that a continuously varying signal can be encoded using binary functions, but the increase in frequency implies that the binary signals encode differing levels of detail. Much energy is carried in the

FIGURE 5.7 Sequency functions

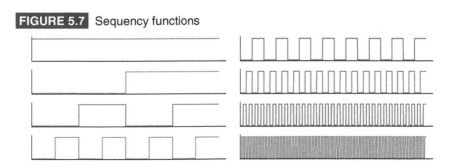

FIGURE 5.8 Applying the 1-D Walsh transform: (a) original signal; (b) Walsh transform

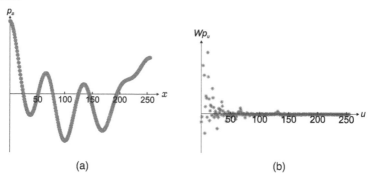

(a)   (b)

FIGURE 5.9 Walsh basis functions, **W**

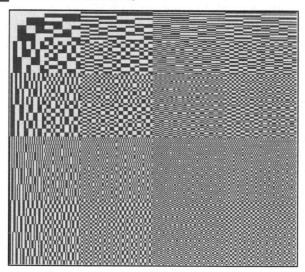

lower-frequency components and the detail is in the higher-frequency components, as in the Fourier transform. Applying the Walsh transform is intrinsically a 'speedy' process since there are no floating-point operations except the additions and the computational cost of a complex (DFT) or real (DCT) sinusoidal operations are avoided. There are fast versions (even in hardware), but we won't go there; here as our interest is the basis rather than the implementation.

The *inverse Walsh transform* is the same process but omits the scale factor (those blooming scale factors! They ever plague us in transform theory). The inverse is

$$p_x = \sum_{u=0}^{N-1} Wp_u \times \prod_{k=0}^{Nb-1} (-1)^{b_k(x)b_{Nb-1-k}(u)} \quad x = 0, 1 \ldots N-1 \tag{5.16}$$

Simplifying the product term

$$p_x = \sum_{u=0}^{N-1} Wp_u \times (-1)^{\sum_{k=0}^{Nb-1} b_k(x)b_{Nb-1-k}(u)} \tag{5.17}$$

Rewriting this using the Walsh transform matrix **W** as

$$p_i = \mathbf{W}^{-1} \times \mathbf{W}p$$

Given that the two functions in Equations 5.14 and 5.17 differ only by a scaling factor

$$\mathbf{W}^{-1} = N \times \mathbf{W}$$

So the reconstruction is

$$\mathbf{p} = N \times \mathbf{W} \times \mathbf{W}p \tag{5.18}$$

The reconstruction of the signal in Figure 5.8(a) from the transform coefficients in Figure 5.8(b) via the Walsh basis functions in Figure 5.9 results in a maximum error of 10^{-16} so we have the usual transform properties of understanding, compression and inversion. The other properties will wait until Section 5.4, as will an investigation of its coding/compression properties. For further study, an excellent text is (was?) available [Beauchamp, 1975] though it seems rather dated (noting that computers then had less computational power than a modern doorbell).

5.2.1.2 The 2-D Walsh transform

The 2-D discrete Walsh transform **WP** of an image **P** is

$$WP_{u,v} = \frac{1}{N} \sum_{y=0}^{N-1} \sum_{x=0}^{N-1} P_{x,y} \times \prod_{k=0}^{Nb-1} (-1)^{b_k(x)b_{Nb-1-k}(u)+b_k(y)b_{Nb-1-k}(v)} \tag{5.19}$$

$$u; v = 0, 1 \ldots N-1$$

with the same simplification of the product term in Equation 5.14. The process is the same as the 1-D version, except that it is in two dimensions and is separable. Using the basis functions of Equation 5.15 we need to image the basis functions so we use a matrix $\mathbf{W} \times \mathbf{W}^T$ product to have

$$\mathbf{WP} = \mathbf{W} \times \mathbf{P} \times \mathbf{W}^T \tag{5.20}$$

which gives a very fast implementation (since **W** can be pre-computed). The transform is shown in comparison with that of the DCT in

FIGURE 5.10 Transforms by DCT and Walsh: (a) image, **P**; (b) DCT, log (**CP**); (c) Walsh, log |**WP**|

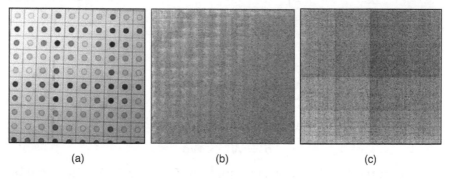

(a)   (b)   (c)

Figure 5.10, for a 256×256 image (the size is fixed by the Matlab implementation) using the same original image so we can compare results. There is less to see than for the other basis, and its properties will be considered for image compression in Section 5.4. The inverse transform results in precisely the same image as **P**, so we shall not display it here.

5.2.2 Walsh–Hadamard transform

The Hadamard transform [Pratt et al., 1969] is also a non-complex transform that changes the order of the sequency components from that of the Walsh transform. For that reason, it is usually known as the Walsh–Hadamard transform, which we shall call it here. It is defined as

$$WHp_u = \frac{1}{N} \sum_{x=0}^{N-1} p_x \times (-1)^{\sum_{k=0}^{Nb-1} b_k(x)b_k(u)} \tag{5.21}$$

Its basis lies in the Hadamard matrix **H**, which is a square matrix that can be computed using a recursive relationship

$$H_{2u} = \begin{bmatrix} H_{u-1} & H_{u-1} \\ H_{u-1} & -H_{u-1} \end{bmatrix} \tag{5.22}$$

So, starting with $H_1 = [1]$, we have the first two functions

$$H_2 = \begin{bmatrix} 1 & 1 \\ 1 & -1 \end{bmatrix} \qquad H_4 = \begin{bmatrix} 1 & 1 & 1 & 1 \\ 1 & -1 & 1 & -1 \\ 1 & 1 & -1 & -1 \\ 1 & -1 & -1 & 1 \end{bmatrix}$$

The Walsh–Hadamard transform matrix **WH** is then derived by normalisation and then re-ordering the rows so as to be in terms of increasing sequency. The ordered four-point transform matrix is then

$$WH_4 = \frac{1}{2}\begin{bmatrix} 1 & 1 & 1 & 1 \\ 1 & 1 & -1 & -1 \\ 1 & -1 & -1 & 1 \\ 1 & -1 & 1 & -1 \end{bmatrix}$$

The 1-D Walsh–Hadamard transform of a sampled signal **p** is then

$$\mathbf{WHp} = \mathbf{WH} \times \mathbf{p} \tag{5.23}$$

The 2-D Walsh–Hadamard transform of an image **P** is

$$\mathbf{WHP} = \mathbf{WH} \times \mathbf{P} \times \mathbf{WH}^{\mathsf{T}} \tag{5.24}$$

We shall not display images here, as they add little to those shown earlier. There is not a great deal of material on the Walsh (Hadamard) transform in the academic literature, with few papers, which are not cited much. Certainly, its compaction/ compression properties are much appreciated. There is also a study of convolution [Robinson, 1972] (logical rather than arithmetic, given the transform basis) and an autocorrelation theorem [Ahmed et al., 1973]. There is always interest in computation, so there is indeed a fast Walsh transform [Shanks, 1969] and a more recent approach, 'which combines the calculation of the Walsh-Hadamard transform (WHT) and the DFT' [Hamood & Boussakta, 2011]. There is a possibility of transfer of Fourier to Walsh as both have (or can be arranged to have) sequency components. Certainly, there is interest in the coding ability of sequency-based functions. Certainly, there is academic interest in the potential for a sequency basis as opposed to more smooth arithmetic functions. But there is no phase, and there is no shift invariance, which perhaps limits more general deployment.

5.3 HARTLEY TRANSFORM

The *Hartley transform* [Hartley, 1942] is a form of the Fourier transform, and without complex arithmetic like the cosine transform. Oddly, though it sounds like a very rational development, the Hartley transform was first invented in 1942 but not rediscovered and then formulated in discrete form until much later in 1983 [Bracewell, 1983] (the Fourier transform had to wait 100 years, so what the heck!). One advantage of the Hartley transform is that the forward and inverse transform is the same operation; a disadvantage is that phase is built into the order of frequency components since it is not readily available as the argument of a complex number.

The definition of the (continuous) Hartley Transform replaces the complex basis by the cas (cos and sine) function defined as

$$\mathrm{cas}\,(t) = \cos(t) + \sin(t) \tag{5.25}$$

The transform is then

$$Hp\left(f\right) = \int_{-\infty}^{\infty} p\left(t\right) \operatorname{cas}\left(2\pi ft\right) dt \tag{5.26}$$

The *inverse Hartley transform* is the same, computing $p\left(t\right)$ from $Hp\left(f\right)$. The *Discrete Hartley Transform* (DHT) similarly replaces the complex exponent by

$$Hp_u = \frac{1}{N} \sum_{x=0}^{N-1} p_x \operatorname{cas}\left(\frac{2\pi}{N} xu\right) \tag{5.27}$$

and this gives the basis of the forward and inverse transforms. The *2-D Hartley transform* **HP** for an $M \times N$ image is computed in forward and in inverse forms as:

$$HP_{u,v} = \frac{1}{\sqrt{MN}} \sum_{x=0}^{N-1} \sum_{y=0}^{M-1} P_{x,y} \operatorname{cas}\left(\frac{2\pi}{M} vy\right) \operatorname{cas}\left(\frac{2\pi}{N} ux\right) \tag{5.28}$$

which is computed using separability. For an $N \times N$ square image, this can be written as

$$HP_{u,v} = \frac{1}{N} \sum_{x=0}^{N-1} \sum_{y=0}^{N-1} P_{x,y} \left(\sin\left(\frac{2\pi}{N}\left(ux + vy\right)\right) + \cos\left(\frac{2\pi}{N}\left(ux - vy\right)\right) \right) \tag{5.29}$$

The inverse Hartley transform is the same process but applied to the transformed image.

$$P_{x,y} = \frac{1}{N} \sum_{u=0}^{N-1} \sum_{v=0}^{N-1} HP_{u,v} \left(\sin\left(\frac{2\pi}{N}\left(ux + vy\right)\right) + \cos\left(\frac{2\pi}{N}\left(ux - vy\right)\right) \right) \tag{5.30}$$

The implementation is then again the same for both the forward and the inverse transforms. Again, a fast implementation is available – the *Fast Hartley Transform* [Bracewell, 1984] (though some suggest that it should be called the Bracewell transform, eponymously). Like the relationship between the DCT and the DFT, it is actually possible to calculate the DFT of a function, FP_u, from its Hartley transform, HP_u. The analysis here is based on 1-D data, for simplicity since the argument extends readily to two dimensions. As for the Fourier transform, by splitting the Hartley transform into its odd and even parts, O_u and E_u, respectively, we obtain:

$$Hp_u = O_u + E_u \tag{5.31}$$

where $E_u = \frac{Hp_u + Hp_{N-u}}{2}$ and $O_u = \frac{Hp_u - Hp_{N-u}}{2}$.

The DFT can then be calculated from the DHT simply by

$$Fp_u = E_u - jO_u \tag{5.32}$$

Conversely, the Hartley transform can be calculated from the Fourier transform by

$$Hp_u = \text{Re}\,(Fp_u) - \text{Im}\,(Fp_u) \qquad (5.33)$$

where Re() and Im() denote the real and the imaginary parts, respectively. This emphasises the intimate relationship between the Fourier and the Hartley transform (and that the FFT can be used for calculation as in the relationship between the DFT and the DCT in Equation 5.8). Code 5.3 illustrates the computation of the Hartley transform in Equation 5.28. There is no cas function in Matlab and it is hardcoded. The same routine performs the forward and the inverse transforms.

Code 5.3 Hartley transform

```
function trans = Hartley(image)
%forward and reverse are the same

[rows,cols]=size(image);
trans(1:rows,1:cols)=0;
%compute transform
for u = 1:cols %address all columns
    for v = 1:rows %address all rows
        sum=0;
        for x = 1:cols %address all columns
            for y = 1:rows %address all rows
                mux=2*pi*(u-1)*(x-1)/cols; % Eq. 5.29
                nvy=2*pi*(v-1)*(y-1)/rows; % no cas in Matlab
                sum=sum+(image(y,x)*(cos(nvy)+sin(nvy))*
                    (cos(mux)+sin(mux)));
            end
        end
        trans(v,u)=sum/sqrt(rows*cols);
    end
end
```

Figure 5.11 shows an example of the transform obtained using Code 5.3. Figure 5.11(a) shows the input image and Figure 5.11(c) shows the result of the transform. Note that the transform in Figure 5.11(c) is the complete transform whereas the Fourier transform in Figure 5.11(b) shows only the magnitude, not the phase. The resulting components have positive and negative values, so to show the results in an image it is necessary to perform some normalisation. The image shown in Figure 5.11(c) shows the log of the absolute value, so it is comparable to the Fourier magnitude, Figure 5.11(b). The transform is symmetrical and this is because both the Fourier and Hartley transforms decompose the image into sinusoidal frequency components.

FIGURE 5.11 Applying the Fourier and the Hartley transforms: (a) image;
(b) DFT, log(|**FP**|); (c) DHT, log(|**HP**|)

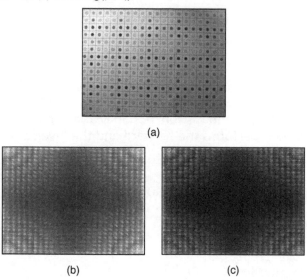

(a)

(b)   (c)

Naturally, as with the DCT, the properties of the Hartley transform mirror those of the Fourier transform. Unfortunately, the Hartley
transform does not have shift invariance, but there are ways to handle this. Also, convolution requires manipulation of the odd and even
parts. The Hartley transform is rather neat and fits the Fourier transform well but the only practical advantage to the Hartley transform is
that the forward and inverse transforms are the same. Now that memory is cheap (and code is written so that it can be read and maintained
rather than just as compact code) then this is not a great advantage. In
terms of compression, the DCT has much better properties, so it is not
surprising that the Hartley transform has found little use. There might
exist a property that has yet to be found, and if that is the case one
could expect a resurgence of interest in the Hartley transform.

5.4   IMAGE COMPRESSION PROPERTIES OF FOURIER, DCT, WALSH AND HARTLEY TRANSFORMS

We shall consider image coding in more detail in the next chapter
(on applications). It is an enormous field, since it is very important.
As we are concerned with principal properties only, the treatment
here will be more rudimentary, lacking any of the finesse of the
modern coding approaches and we shall consider only what happens when a fixed proportion of components exceeding some chosen value, *threshold*, is retained before reconstruction using an inverse

transformation (thus terming it compression). We shall perform a transform and then choose a threshold so as to retain 20% of the largest components (discarding those of least magnitude), as

$$
thresholded\_transform_{u,v} = \begin{cases} 0 & transform_{u,v} < threshold \\ transform_{u,v} & transform_{u,v} \geq threshold \end{cases}
$$

(5.34)

The thresholding process removes some components and we shall use some measures to compare the reconstruction performance. The measure is the number of components remaining, expressed as a percentage of image size, here 20%. After some of the transform components have been deleted, the image reconstructed from them will differ from the original version. In practice the appearance of any image is a question of perceptibility, whereas to measure the difference in reconstruction we shall use the Mean Absolute Error, *MAE* between the original and the image reconstructed from fewer components, where

$$
MAE = \frac{1}{N^2} \sum_{x=0}^{N-1} \sum_{y=0}^{N-1} \left| original_{x,y} - reconstructed_{x,y} \right|
$$

(5.35)

We shall also consider the maximum absolute difference, *max_diff*, which is normalised by the maximum value of the original image to be expressed as a percentage.

$$
max\_diff = 100 \times max \left| original_{x,y} - reconstructed_{x,y} \right| / max\left(original \right)
$$

(5.36)

The performance of the different algorithms on one image is shown in Figure 5.12 and Table 5.2. The transforms were thresholded at a value that selected 20% of the largest components (the large apparent difference between the thresholds is because of the different scaling factors used). There is no structure to this choice, and structure/ informed choice could clearly improve matters. Noting that there appears to be little difference between most of the reconstructed images, Figures 5.12(b)–(e), the performance of the transforms is actually different and the DCT appears to offer the best performance by these error metrics. This is reflected in the detail of the camel riders which appears clearer by the DCT and Hartley transforms, it is very unclear in the Fourier reconstruction whereas it has been zapped in the Walsh transform – noting that this inverse Walsh transform can reconstruct exactly when coding has not been used. That the DCT is best (via optimised coding) is generally perceived to be the case, over considerably many more different images than we have considered here. Note that refinement of technique can apply equally to all the transforms so this analysis

FIGURE 5.12 Images compressed by thresholding transforms: (a) original image; (b) reconstruction and detail by thresholded FFT; (c) reconstruction and detail by thresholded DCT; (d) reconstruction and detail by thresholded Hartley; (e) reconstruction and detail by thresholded Walsh-Hadamard

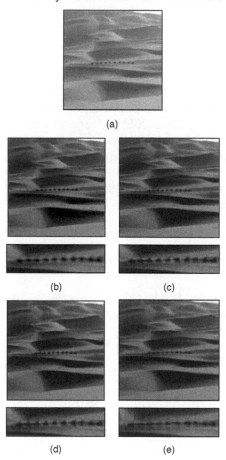

TABLE 5.2 Image compression performance by thresholding transform data.

Transform	Threshold	Proportion	MAE	max_diff
DFT (fft2)	3663	20%	4.02	18.5%
DCT (2-D Code 5.1)	1664	20%	3.49	16.8%
Hartley (Code 5.3)	13.7	20%	3.68	17.0%
Walsh-Hadamard (Equation 5.24)	5885	20%	6.42	32.7%

is baseline only, and for one image only. The transforms offer different insights and have different advantages. Coding might be the main application, but there are many others.

5.5 LAPLACE, MELLIN AND FOURIER MELLIN

We should admit here that the target of this section is the 2-D Fourier-Mellin transform which can be used to find objects in images, whatever their position, size and rotation (shapes that have undergone an affine transform). So it is good stuff, and more details will come later. To get there requires a bit of a diversion since we shall consider some transforms which largely reside in signal processing, and that is not the target of this text. We do however have to include them, as otherwise Fourier–Mellin is a bit of a jump, and they fit with other material. The first, Laplace, has already been mentioned in Section 2.6.1.

5.5.1 Laplace and Mellin transforms

5.5.1.1 Laplace transform and basic systems analysis

The *Laplace transform Lf*() can be defined for a function of time $f(t)$ using a complex variable $s = \sigma + j\omega$, evaluated overall time as

$$Lf(s) = \int_{-\infty}^{\infty} f(t)\, e^{-st} dt \qquad (5.37)$$

This is the two-sided definition and an alternative is the one-sided version, which is evaluated over the interval between (just above) 0 and ∞. There is slight irony including Laplace in a book on Fourier, as they had their differences as we shall find in the history, Section 7.1.2. The inversion formula is similar to Fourier, but it is not in terms of a real variable but it is in terms of a complex one, so it becomes rather nasty indeed. The function is regained via a contour integration as

$$f(t) = \frac{1}{j2\pi} \int_{c-j\infty}^{c+j\infty} Lf(s)\, e^{st} ds \qquad (5.38)$$

for a constant c. This looks rather complicated, and contour integration certainly is. But in practice the process is much simpler since there are established tables of previously calculated Laplace transform pairs, which can be used to substitute into factorised equations (partial fractions are happily rolled out to resolve the transform components). Since they are transform pairs, the time domain functions can be evaluated easily. Also, the manipulation in the frequency domain is analytically attractive since the Laplace transform has very similar properties to the Fourier transform in terms of addition, superposition, shift, scaling, convolution, etc. It is a very handy tool for signal processing and it can be used, for example, to solve easily for the RC filtering networks in Section 2.6.1. First, let us see some transform pairs and properties, shown in Table 5.3. Some of the major properties (that we shall use) include addition, shift, convolution and differentiation, and these are given analytically. Given that – with apologies to real mathematicians – the Laplace operator s can be considered as a multiplicative operator akin with differentiation (Property 4), we can note that the pairs for the first-order terms in s are exponential functions, and the pairs for the higher-order terms in s^2 include sinusoidal functions of time, which is to be expected as higher-order differentiation emphasises change.

TABLE 5.3 Example Laplace transform properties and pairs.

Laplace transform properties		
Property	**Time $f(t)$**	**Laplace $Lf(s)$**
1 Addition	$f_1(t) + f_2(t)$	$Lf_1(s) + Lf_2(s)$
2 Shift	$f(t - \tau)$	$e^{-s\tau} Lf(s)$
3 Convolution	$f_1(t) * f_2(t)$	$Lf_1(s) \times Lf_2(s)$
4 Differentiation	$f' = df(t)/dt$	$s \times Lf(s)$
Laplace transform pairs		
	Time	**Frequency**
A	$e^{-\alpha t}$	$\dfrac{1}{s + \alpha}$
B	$te^{-\alpha t}$	$\dfrac{1}{(s + \alpha)^2}$
C	$\cos(\omega t)$	$\dfrac{s}{(s^2 + \omega^2)}$
D	$e^{-\alpha t}\cos(\omega t)$	$\dfrac{s + \alpha}{((s + \alpha)^2 + \omega^2)}$

By the Laplace transform the current/voltage relationship in the RC network in Sections 2.3.1 and 2.6.1 is, by the differentiation, Property 4

$$i = C \times sv_C$$

which can be substituted into the equation derived for the voltage across the resistor as

$$v_i - v_C = RCsv_C$$

By rearrangement this gives the transfer function

$$\frac{v_C}{v_i} = \frac{1}{1 + RCs}$$

From Table 5.3, Transform Pair A, for an input A, the solution is

$$v_C(t) = Ae^{-t/RC}$$

When the input switches off, this is the exponential decay associated with the capacitor discharging through the resistor. This is much easier by Laplace transform analysis than using differential equations. Differing values of the time constant RC change the rate at which the output decays and it is a low-pass filter since for small values of RC when the input switches off (or on), the capacitor cannot discharge (or charge) fast enough when the input signal changes quickly, thus blocking the high-frequency signals. As we saw earlier, for frequency domain analysis we omit the d.c. terms by substituting $s = j\omega$. As we have often said, however tempting it is to digress there (applying maths is the best bit of maths), signal processing is beyond the scope here and is the focus of many (much larger) books.

5.5.1.2 Mellin transform for scale invariance

The *Mellin transform* [Bracewell, 1986], named after the Finnish mathematician Hjalmar Mellin, achieves scale invariance by compressing the time axis to a particular interval, in a particular scale. This allows for the comparison of two signals that exist over different intervals and at different scales. An excellent coverage is available online [Bertrand et al., 2000] and is explicit concerning the complexity of this transform. Some of the difficulty is analytic, and some concerns implementation. Given these complexities, we shall skate quickly here. The motivation for the transform is clear, but there are alternative solutions now available in some domains – perhaps even motivated by the complexity here. The transform is defined as

$$FMf(s) = \int_0^\infty f(x)\,x^{s-1}dx \tag{5.39}$$

noting that the time index is now changed to x, which is implicit in the Mellin transform. To understand the process, we need to determine the similarity to the Laplace transform. When we make the substitution

$$x = e^{-t}$$

By differentiation this gives $dx = -e^{-t}dt$

and by raising the power $x^{s-1} = e^{-t(s-1)}$

By substitution into Equation 5.39 we have

$$FMf(s) = \int_0^1 f\left(e^{-t}\right)e^{-st}dt \tag{5.40}$$

The time axis has now been compressed logarithmically as a function of e^{-t} rather than as a function of t, which is the reason for the change

FIGURE 5.13 Time axis compression in the Mellin transform: (a) original signal; (b) with time axis logarithmically compressed (and inverted)

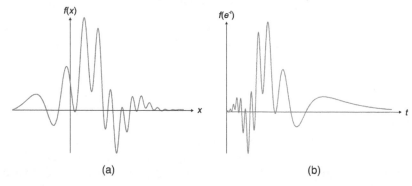

(a)   (b)

of variable in its definition, Equation 5.39. Also, the time axis has been compressed. This is illustrated in Figure 5.13 for an original signal and with the time axis compressed. If the original signal was to occur over a different time period, then the two signals would be scaled so as to appear the same in the transform.

The *inverse Mellin transform* uses a contour integral

$$f(s) = \frac{1}{j2\pi} \int_{c-\infty}^{c+\infty} FMf(s)x^{-s}ds \qquad (5.41)$$

where c is a constant. 'This … is not always the simplest. In some cases, however, the integral can be computed by the method of residues' [Bertrand et al., 2000]. Ouch. Also, to deploy the Mellin transform for detection, we need an inversion of the convolution/correlation process, and even convolution is difficult.

To translate the approach to sampled signals sounds eminently suitable but raises some difficulties. The *discrete Mellin transform* does not appear in Wikipedia but it certainly exists [Bertrand et al., 2000], which is unsurprising since in speech processing there is a desire to match speech irrespective of when it started and finished and at what speed it was uttered. That is what the Mellin transform affords. There is a certain amount of fog here, and the mathematical definitions are very complex indeed (even including 'dilatocycled' functions no less). Perhaps this and the difficulty in implementation have led to little use and it is more common now to achieve pattern matching using the Dynamic Time Warping (DTW) algorithm, which involves nonlinear warping of the axes of the data, rather than using the frequency domain and is thus out of scope here. In speech recognition, feature extraction using Mel-Frequency Cepstral Coefficients (MFCCs) uses the frequency domain, as we shall find in the next chapter.

The 2-D discrete Mellin transform [Altman & Reitbock, 1984] essentially scales the spatial coordinates of the image using an exponential function, in the manner of Equation 5.40. A point is then moved to a position given by a logarithmic function of its original coordinates. To achieve template matching, the transform of the scaled image is then multiplied by the transform of the template. The maximum again indicates the best match between the transformed data and the image. This can be considered to be equivalent to a change of variable. The logarithmic mapping ensures that scaling (multiplication) becomes addition. By the logarithmic mapping, the problem of scale invariance becomes a problem of finding the position of a match. Naturally, this involves many complications that include interpolation since images are discrete rather than continuous.

5.5.2 Fourier–Mellin transform

The Mellin transform provides scale-invariant analysis. For scale- and position-invariance, the Mellin transform is combined with the Fourier transform, to give the *Fourier–Mellin* transform. The Fourier–Mellin transform has many disadvantages in a digital implementation, because of the problems in spatial resolution though there are approaches to reduce these problems [Altmann & Reitbock, 1984]. The interest in image analysis is to register images (render them into the same presentation to allow comparison) and what appears to be the earliest work on scale- and rotation-invariant image matching [Castro, 1987] does not even mention the Mellin transform. The usual approach to *log-polar mappings* for image analysis [Zokai & Wolberg, 2005] includes the following steps:

1. Form the FFT of the image to form $FP_{u,v}$;
2. Perform a complex log mapping where u, v are mapped using polar notation $re^{j\theta}$ to form $FM_{\ln r, \theta}$ where r, θ are the distance and direction to the centre, mapping radial lines to Cartesian space;
3. Sample **FM** uniformly along the logarithmic scale and using polar coordinates;
4. Take an inverse FFT of the resulting image.

The first stage gives position invariant analysis by the shift-invariance property, the second stage is a discrete version of the Mellin transform allowing scale invariance and the third stage allows for rotation-invariance. Thus, the only interest here for Fourier becomes the shift-invariance property and this is well known, so we shall just supply an image that shows the registration of one image to another, by log-polar mapping, shown in Figure 5.14. As image registration is a large field indeed (in hospitals, there are Professors of Registration) there are many and more advanced techniques than this [Maintz &

FIGURE 5.14 Applying a log-polar mapping for image registration [Zokai & Wolberg, 2005]: (a) template; (b) finding the template in an image

 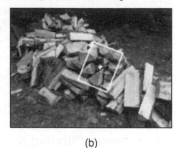

(a)   (b)

Viergever, 1998; Viergever et al., 2016]. Naturally, the approach could be applied in one dimension, omitting rotation.

The Mellin transform appears to be much better suited to an optical implementation [Casasent & Psaltis, 1977], where continuous functions are available, rather than to discrete image analysis. A further difficulty with the Mellin transform is that its result is independent of the *form factor* of the template. Accordingly, a rectangle and a square appear to be the same to this transform. This implies a loss of information since the form factor can indicate that an object has been imaged from an oblique angle.

5.6 z-TRANSFORM

Given that we have just breezed through continuous signal processing and the Laplace transform, we shall now note that there is a discrete equivalent. For sampled data, the operator is known as the *z-transform*, and z^{-1} represents a time delay of one sampling instant. It is actually an operator, so we can perform differencing (from Section 3.3.7). The difference between adjacent samples is

$$\mathbf{p}' \equiv \mathbf{p}[m] - \mathbf{p}[m-1]$$

Using z-transform notation this is then

$$\mathbf{p}' \equiv \mathbf{p}[m] - z^{-1}\mathbf{p}[m] = \mathbf{p}[m]\left(1 - z^{-1}\right) \qquad (5.42)$$

Also, a digital filter is depicted in Figure 3.17. Using z-transform notation

$$\text{output} = p_N + z^{-1}p_{N-1} + z^{-2}p_{N-2} + z^{-6}p_{N-6} + z^{-7}p_{N-7} + z^{-8}p_{N-8} \qquad (5.43)$$

and by factorisation, the output p_O is calculated from the input p_i as

$$p_O = p_i\left(1 + z^{-1} + z^{-2}\right)\left(1 + z^{-6}\right) \qquad (5.44)$$

which results in a discrete transfer function H (z)

$$H(z) = \left(1 + z^{-1} + z^{-2}\right)\left(1 + z^{-6}\right) \tag{5.45}$$

The transfer function can be used to calculate the response of the system to a selected (discrete) input signal. Given that there is a standard substitution (by conformal mapping, evaluated along the frequency axis) $z^{-1} = e^{-j\omega t}$ that transforms from the time domain (z) to the frequency domain (ω), then we can substitute this as

$$H(j\omega) = \left(1 + e^{-j\omega} + e^{-2j\omega}\right)\left(1 + e^{-6j\omega}\right) \tag{5.46}$$

from which we can calculate the frequency response of the digital filter. We shall not do that here since by Equation 5.46 the frequency response is formed from the concatenation of two types of averaging (and to be frank, it looks awful – 'it conveys little' as academics say).

5.7 WAVELETS

5.7.1 Filter banks and signal analysis

One approach to analysing signals is to construct *filter banks,* which are sets of band-pass filters that cover the signal's spectrum. This is shown in Figure 5.15 where each of the filters has an equal bandwidth. In other cases this might differ, for example, the human sensitivity to frequency decreases with an increase in frequency. Therefore, to approximate human hearing the bandwidths within the filter bank become wider with increase in frequency. A filter bank could be used, say, to equalise the performance of a microphone. Since microphones do not respond equally across the audio spectrum, and it can be desirable to do so for music recording, then a filter bank could be deployed so as to ensure that the microphone gives the same response to any acoustic signal.

A filter bank requires implementation, as in Figure 5.16, where details of the signal processing, which conditions the signal in different ways, have not been shown. In this way, we can access the different frequencies that make up a signal. This can be used in coding, as it allows for a more efficient *sub-band coding* scheme. It can also be used for detection, since the filters can be tuned to the ranges of frequencies they are to be used to detect.

The sensitivity to frequency in the filter banks and transforms so far is global: we can access a frequency within a signal whenever it occurs and that is the global property of the Fourier transform. There is a class of approaches that are sensitive not only to frequency but also to time: the techniques can detect frequencies that occur at particular times. These are wavelets, coming next.

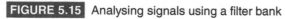

FIGURE 5.15   Analysing signals using a filter bank

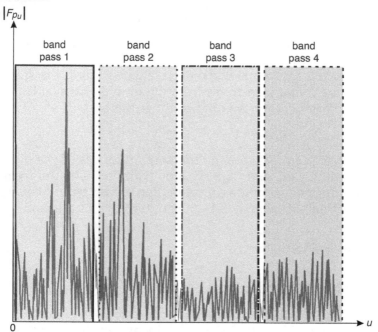

FIGURE 5.16   Filter bank analysis system

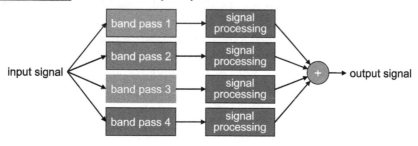

5.7.2   Gabor wavelets

Previously we have covered windowing, which is used to select particular instances of signals. We have also mentioned the Short-Time Fourier transform, which uses windowing to localise frequency analysis. The analysis of both approaches embodies the convolution theorem. Their restriction is that the duration of the windowing function is fixed and there is a trade-off in the choice of the window size. A longer window affords more local support in the time domain and is more localised in the frequency domain. Given the scaling property of the Fourier transform, we can change the window length to include *scale* in the analysis. In images, wavelets are sensitive to frequency and to space and can detect frequencies that occur in different parts of an image and at different scales. An alternative view is that the wavelets represent windowing functions of differing size (duration) matched to the signals that are being analysed. These wavelet approaches are a more recent approach to signal processing than the Fourier transform, being introduced in the 1990s [Daubechies, 2009].

Wavelets' main advantage is that they allow sensitivity to frequency/time and frequency/space and allow for multi-resolution analysis (analysis at different scales, or resolutions). Another view is that wavelets *localise* signals both in the time and in the frequency domains. Simultaneous decimation allows us to describe an image in terms of frequency which occurs at a position, as opposed to an ability to measure frequency content across the whole image. Clearly this gives us a greater descriptional power than the Fourier transform, which can be used to good effect.

Denis Gabor is often more renowned for inventing holograms but was a major player in the uncertainty theorem which is why it is often called the Gabor trade-off. According to my PhD supervisor, he was regarded as a bit of a crank at Imperial – a 3D picture from light, ooh I say. Fame can come late in academe.

First though, we need a basis function, so that we can decompose a signal. The basis functions in the Fourier transform are sinusoidal waveforms at different frequencies. The function of the Fourier transform is to convolve these sinusoids with a signal to determine how much of each is present. The *Gabor wavelet* is a sine wave modulated by a Gaussian envelope, which is effective for analysis given its properties shown in the uncertainty theorem. A 1-D Gabor wavelet *gw* is given by

$$gw\left(t; t_0, \Omega, \sigma\right) = ke^{-j\Omega t}e^{-\left(\frac{t-t_0}{2\sigma}\right)^2} \tag{5.47}$$

where $\Omega = 2\pi f_0$ is the modulating frequency, t_0 dictates position and σ controls the width of the Gaussian envelope that embraces the oscillating signal and k is for scale. The duration of the envelope dictates the length of local support for chosen frequency. This is very similar to the later *Morlet transform*, which is used in audio analysis [Goupillaud et al., 1984]. Equation 5.47 can be considered as a *mother wavelet* when constructing a *wavelet family* in which the *child wavelets* have a structural relationship with the mother wavelet. An example Gabor wavelet is shown in Figure 5.17, which shows the real and the imaginary parts (and their Gaussian envelope). Increasing the value of Ω increases the frequency content within the envelope, whereas increasing the value of σ spreads the envelope without affecting the

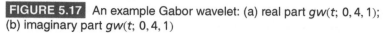

FIGURE 5.17 An example Gabor wavelet: (a) real part $gw(t;\,0,4,1)$;
(b) imaginary part $gw(t;\,0,4,1)$

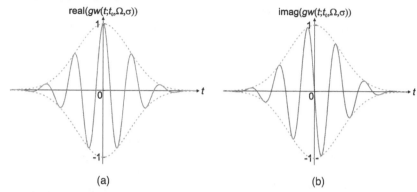

frequency. So why does this allow simultaneous analysis of time and frequency? Given that this function is the one to be convolved with the test data, then we can compare it with the Fourier transform. In fact, if we remove the rightmost term of Equation 5.47, we return to the sinusoidal basis function of the Fourier transform as in Equation 2.1. Accordingly, we can return to the Fourier transform by setting σ to be very large. Alternatively, setting Ω to zero removes the frequency information, returning to a Gaussian waveform. Since we operate in between these extremes, we obtain position and frequency information simultaneously.

The effect of changing the parameters is shown in Figure 5.18, given the wavelet of Figure 5.17 as the mother wavelet. We shall consider only the even part of the wavelet. In Figure 5.18(a), changing Ω whilst not changing σ changes the baseline frequency, and not the envelope. With this envelope, then the number of visible peaks is related to the value of Ω and the sinusoidal waveforms fit inside the Gaussian envelope. Changing σ changes the spread of the Gaussian envelope. In Figure 5.18(b), when Ω is constant the frequency content is the same (the three curves inside the Gaussian envelopes are of the same frequency), whereas the spread controlled by σ changes.

Actually, an infinite class of wavelets exists, which can be used as an expansion basis in signal decimation. One approach [Daugman, 1988] has generalised the Gabor function to a 2-D form aimed to be optimal in terms of spatial and spectral resolution. These 2-D Gabor wavelets are given by

$$gw2D\left(x,y;x_0,y_0,\Omega,\sigma,\theta\right) = ke^{-j\Omega((x-x_0)\cos\theta+(y-y_0)\sin\theta)}e^{-\left(\frac{(x-x_0)^2+(y-y_0)^2}{2\sigma^2}\right)}$$

$$\tag{5.48}$$

FIGURE 5.18 Changing the parameters of a Gabor wavelet: (a) changing the frequency Ω, envelope constant $\sigma = 1$; (b) changing the envelope σ, frequency constant $\Omega = 4$

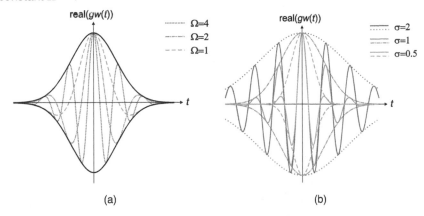

(a)   (b)

FIGURE 5.19 A 2-D Gabor wavelet (a) real part; (b) imaginary part

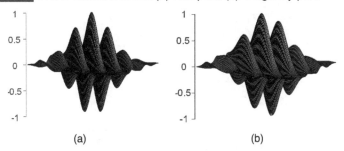

(a)   (b)

where x_0, y_0 control position, $\Omega = 2\pi f_0$ controls the frequency of modulation along the axes and θ controls the direction (orientation) of the wavelet (as implicit in a 2-D system). Naturally, the shape of the area imposed by the 2-D Gaussian function could be elliptical if different variances were allowed along the x and y axes (the frequency can also be modulated differently along each axis). We shall demonstrate the application of wavelets analysing images so we can see the operation and the results. Figure 5.19 shows an example 2-D Gabor wavelet, which reveals that the real and imaginary parts are even and odd functions, respectively. Again, different values for Ω and σ control the frequency and envelope's spread, respectively, and the extra parameter θ controls rotation.

The function of the wavelet transform is to determine where and how each wavelet specified by the range of values for each of the free parameters occurs in the image. Clearly, there is a wide choice that depends on application. An example wavelet analysis is given in Figure 5.20 from which a filter bank of 64 × 64 images is derived

FIGURE 5.20 Illustrating application of a Gabor wavelet filter bank: (a) mother wavelet (surface); (b) mother wavelet (image); (c) target image; (d) Ω A, σ A, θ A; (e) Ω A, σ C, θ A; (f) Ω A, σ A, θ D; (g) Ω C, σ A, θ A; (h) Ω A, σ B, θ B; (i) Ω B, σ A, θ B; (j) Ω B, σ D, θ B; (k) Ω C σ C, θ C; (l) convolution of (i); (m) convolution of (k)

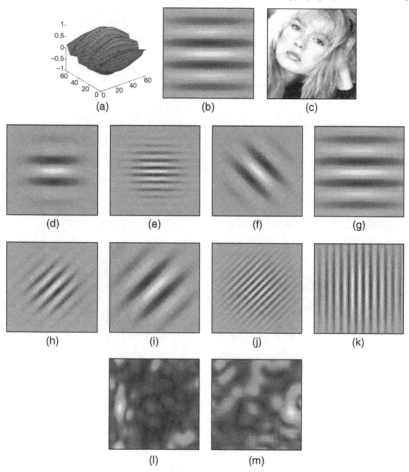

from the mother wavelet in Figure 5.20(a) and (b) and applied to the 256×256 image in Figure 5.20(c). There are four values for each of the parameters Ω, σ and rotation θ. These values are given labels between A and D, which represent integer multiples of the parameters used in the mother wavelet. Naturally the choice of ranges for the variations in the parameters in the wavelet family depends on application (and our application is just to show how the bloody thing works). Four values of rotation are used: rotation A is 0°, rotation B is by 45°, rotation C is by 90° and rotation D is by 135°. Others prefer intervals of 60°, though many others can be used. Altogether, there are 64 wavelets in the filter bank (four rotations, four scales and four frequencies), and we just show a selection of these. For the child wavelet shown in Figure 5.20(d): Figure 5.20(e) shows change in the frequency Ω; Figure 5.20(f) shows change in the rotation; and Figure 5.20(g) shows change in the scale σ. A variety of other combinations are shown in Figures 5.20(h)–(k). The 64 wavelets were zero-padded and convolved with the image (using the FFT via the convolution theorem, noting that with these vector–matrix operations Matlab is doing what it was designed for and it works blindingly fast). Application begs a question of what to do with the 64 resulting images. One can average the results, perform a weighted average or just take the maximum. Here, we show two examples of the results for two of the wavelets: Figure 5.20(l) shows the result of convolving the diagonal lower frequencies of Figure 5.20(i) resulting in a peak near the lady's right cheek; Figure 5.20(m) shows the result of convolving the high frequencies of Figure 5.20(k) resulting in a peak near the hair.

Not that it's anything to do with Gabor, but the lady in Figure 5.20 had an interesting career: she started as a lady of the night, went on to be a pop star and then wrote a book about it. In her career she had a lot of positions, especially at the start.

Here, the Gabor wavelet parameters have been chosen in such a way as to select face features: the cheek and hair have come out well in these results. These features are where there is local frequency content with orientation according to the head's inclination. Naturally, these are not the only features with these properties, the eyes, mouth and even the cuff of the sleeve can be highlighted too! But this does show the Gabor wavelet's ability to select and analyse localised variation in image intensity. Let us make the key point 9.

Key point 9

Wavelets allow sensitivity to frequency/time and frequency/space and allow for multi-resolution time-scale analysis.

The conditions under which a set of continuous Gabor wavelets will provide a complete representation of any image (i.e. that any image can be reconstructed) were developed later. However, the theory is naturally very powerful, since it accommodates frequency and position simultaneously, and further it facilitates multi-resolution analysis – the analysis is then sensitive to scale, which is advantageous since objects that are far from the camera appear smaller than those that are close. We shall find wavelets again, when processing images to find low-level features. Amongst applications of Gabor wavelets, we can find measurement of iris texture to give a very powerful biometric-based security system [Daugman, 1993] and face feature extraction for automatic face recognition [Lades et al., 1993]. Wavelets continued to develop [Daubechies, 2009] and have found applications in image texture analysis [Laine & Fan, 1993], in coding [da Silva & Ghanbari, 1996] and in image restoration [Banham & Katsaggelos, 1996]. Unfortunately, the discrete wavelet transform is not shift invariant, though there are approaches aimed to remedy this (see e.g. [Coifman & Donoho, 1995]). In fact, the *Dual Tree Complex Wavelet Transform* is a more recent approach that offers 'important additional properties: It is nearly shift invariant and directionally selective in two and higher dimensions' and it is beautifully described [Selesnick et al., 2005]. An excellent place to start on localised frequency analysis with orthonormal basis functions is with one of its leaders: try [Mallat, 2008]. As interesting and important as it is, we shall not study it further and just note that there is an important class of transforms that combine spatial and spectral sensitivity, and it is likely that this importance will continue to grow.

5.8   SUMMARY

Some works actually manage a veneer of criticism about Fourier's use of sine and cosine waves as they are too general. That is simply daft, since it was an excellent choice and laid the foundations for many later transforms, and these have been covered in this chapter. Essentially, the restrictions associated with using a complex exponential can be handled in two ways. The first is to separate out the components of the complex exponential, and handle them separately to give unitary transformation as in the Cosine, Sine and Hartley transforms. Then we can recognise that sinusoids are simply a regular differentiable periodic signal that can be exchanged for other functions, here for the Walsh and the Walsh–Hadamard discrete basis. The limitation of the Fourier scaling theorem can be obliterated by rescaling the time axis in the Mellin transform, though its implementation is complex. Shift invariance can be included within the Fourier–Mellin transform, but in application reasonable approximations appear to be preferred. We have also considered operators in signal processing, because transforms are important in systems analysis (and because we left them out earlier). The Laplace transform is used for continuous systems and the z-transform is used for discrete systems analysis. Finally, the Fourier approach is sensate only to global frequency. Wavelets introduce sensitivity to time/ space concurrently with frequency, and give a richer basis of signal analysis and offer a richer basis for windowing. As we have now covered all the main theory and approaches consistent with an introductory text we shall move on to applications, coming next in Chapter 6.

Applications of the Fourier Transform

6.1 OVERVIEW

We shall now describe a variety of applications where the Fourier transform, or one of its variants, has a critical advantage that allows insight or performance to be achieved. We have alluded to applications previously, aiming to clarify the basis of technique, and this is extended here. The variety here is quite wide, and we assure you dear reader that the selection is not whimsical caprice but a salmagundi of applications selected to show advantage. It is a vignette, because Fourier can be used in many areas of applied science. Many of the applications here are image-based, not because it is Mark's research area but because we take advantage of the thousand words that each image affords. Some of the image-based applications are indeed from Mark's research team (and are covered in more detail elsewhere [Nixon & Aguado, 2019]), which partly motivated the genesis of this book. The Fourier transform (FT, and the fast Fourier transform FFT) allow for new insight and new performance, and that is what we describe here whilst covering the properties and advantages of transform-based analysis.

This chapter shows how the Fourier transform can be used and so is structured largely to follow the whole book. First, there is 1-D analysis, followed by 2-D and then variants of the Fourier transform. The properties follow in the same order as they were covered in Chapters 2 and 3, using images, moving on to Chapters 4 and 5. Because it is illustrative, it is maximal images and minimal maths, sometimes just the principal equation. This is the only part of the book that is not inherently reproducible, without the implementations given for the rest of the book. Links to implementation are given where available, though much of the material precedes the notion of reproducibility (some of it even preceding Matlab).

It is worth noting that the performance attributes of the variants of the Fourier transform motivated their development. The basis functions of the Fourier transform are very general and that gives the Fourier transform some of its properties, and power, but can restrict its application capability, and its variants can offer improved performance. This is seen especially in coding but also in other applications. Many of the earlier studies that originally used the Fourier transform later developed new techniques for analysis and implementation, some of which were described in Chapter 5. As we shall find in some of the applications, beyond the use of the FFT, Fourier gives the original insight (and faster computation) but a more specialised approach sometimes delivers the goods.

6.2 FOURIER TRANSFORMS

6.2.1 The continuous Fourier transform and Fourier optics

It is actually possible to form the Fourier transform optically, using transmitted light thereby giving a continuous Fourier transform with very fast processing. It is a large subject, and there are even books dedicated to it [e.g. Goodman, 2005]. Essentially, the Fourier transform can be formed using a lens, as shown in Figure 6.1, to give an *optical Fourier transform*. The process employs a light source and a collimator, which produces a parallel beam from it (the light source can be a laser so it is monochromatic). This is used to irradiate a transparency of an image, such as an old slide photograph, $P(x, y)$ placed in the focal plane of a lens. It can be shown (i.e. this stuff is innately complex and we'll avoid it) that the image $I(x, y)$, as formed by a diffraction process and in the focal plane of the lens, is derived from the Fourier transform $FP(u, v)$ [Klein & Furtak, 1986] as

$$I(x, y) = \mathcal{F}(FP(u, v)) = \frac{j\mathbf{P_0}}{\lambda f} e^{jk_1} e^{j\frac{k_2 r^2}{2f^2}(S-f)} FP(u, v) \tag{6.1}$$

where f is the focal length of the lens, $\mathbf{P_0}$ is the illumination intensity and λ is its wavelength, $u = \frac{-x}{\lambda f}$ and $v = \frac{-y}{\lambda f}$ are the spatial frequencies, k_1 and k_2 are determined by the physical arrangement of the imaging system, and the coordinates in the focal plane are

$$r^2 = (u - u_0)^2 + (v - v_0)^2$$

which derive from the Fresnel approximation. The transform shown within Figure 6.1 was not calculated optically (it has been pinched from Chapter 4 as have the other images here). It does not matter precisely where the image is placed, as long as it is within the focal plane, by the shift-invariance theorem.

FIGURE 6.1 Optical Fourier transform

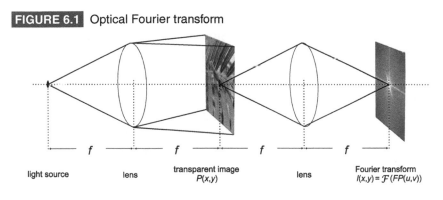

light source   lens   transparent image   lens   Fourier transform
                                          P(x,y)   I(x,y) = $\mathcal{F}$(FP(u,v))

FIGURE 6.2 Optical filtering

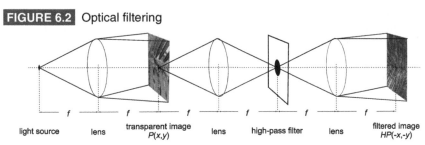

light source   lens   transparent image   lens   high-pass filter   lens   filtered image
                                P(x,y)   HP(-x,-y)

Now that we know the lens can be arranged to lead to a Fourier transform, we can process the transform and then use another lens to reconstruct the image by inverse transformation. The optics leads to inversion and for an image $P(x, y)$ we reconstruct an image $HP(-x, -y)$. Because we have a transform in the middle of the sequence of operations, we can process it. Here we show the introduction of a high-pass filter in Figure 6.2 blocking the low frequencies near the centre of the transform, though we could have equally shown low-pass, band-pass or even convolution/correlation.

The process uses transmission not reflection, so it is not possible to form the Fourier transform of a printed image in this way. That would make the optical process highly attractive, but it is not available (yet?). This material is optics and though is an excellent illustration of a continuous Fourier transform and a natural delight to physicists (it leads to Fourier-based holography), we shall not irrupt a theoretical odyssey into a niche technical area, and just move on to another application.

6.2.2 Magnitude and phase, and beamforming

Detecting submarines has a history as long as submarines themselves and has spawned much technical innovation. Naturally, approaches are based on reflection of sound but mere reflection indicates existence rather than position. It is possible to deploy a *sensor array* and then aim to maximise reception in one position and to minimise interference from other positions. One deployment of this is in towed sonar arrays to detect underwater submarines, Figure 6.3, and is in general use called *beamforming* [Van Veen & Buckley, 1988]. The ability to focus on position is not just limited to sonar and submarines but also deployed in applications including phased-array and synthetic-aperture radar, satellite communications and medical imaging.

Figure 6.4 shows the response of a sonar array aimed to estimate the Direction of Arrival (DOA) of a target, here at an orientation of around 20°. Adverse conditions such as other targets and that the sonar array will not deploy in a straight line as well as the laminar structure of water layers will confound reception, though the known capability of the Fourier transform can help to mitigate these issues.

Planar and spatial array structures are shown in Figure 6.5. The output of the planar array in Figure 6.5(a) is

$$o(k) = \sum_{y=1}^{R} w_y s_y(k) \qquad (6.2)$$

FIGURE 6.3 Towed sonar array

FIGURE 6.4 Beamformer response [Van Veen & Buckley, 1988]

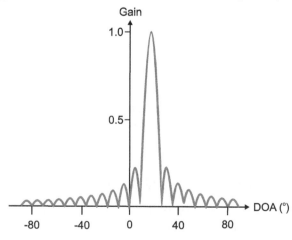

FIGURE 6.5 Sensor arrays: (a) planar; (b) spatial

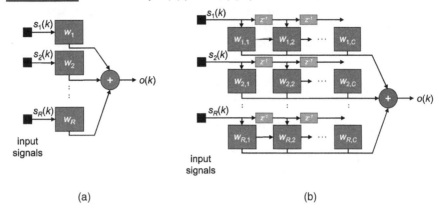

(a)   (b)

and the output of the spatial array (using z-transform notation, Section 5.6, to represent time delay) in Figure 6.5(b) is given by

$$o\left(k\right) = \sum_{y=1}^{R} \sum_{x=1}^{C} w_{x,y} s_y\left(k\right) z^{-(x-1)} \tag{6.3}$$

The frequency response $r(\omega)$ of a planar line where the first element has zero phase and there is a delay of T seconds between successive stages is

$$r(\omega) = \sum_{x=1}^{C} w_x e^{-j\omega T(x-1)} \tag{6.4}$$

When this is extended to two dimensions in a spatial array of the form in Figure 6.5(b), and where the first-row element has no delay, successive elements in each row have a delay $\Delta_y(\theta)$ from the first sensor to sensor y. The response is affected by direction θ (the DOA; range has not been considered here) as

$$o(k) = e^{j\omega k} \sum_{y=1}^{R} \sum_{x=1}^{C} w_{x,y} e^{-j\omega(\Delta_y(\theta)+x-1)} = e^{j\omega k} r(\theta, \omega) \tag{6.5}$$

The non-ideal characteristics of the sensing situation can be accommodated within this structure. The beam pattern is the magnitude of the response, and parameters need to be chosen so as to give the desired response function in Figure 6.4. Essentially, this is using a spatial arrangement introducing phase and that provides the discrimination capability, and there is a wealth of material in analysis and deployment of these complex sensing systems. There is a rich literature in this area, describing a variety of applications [e.g. Benesty et al., 2018]. In plenoptic cameras [Wu et al., 2017], the use of phased arrays of sensors allows the ability to focus not only on space but also on distance. In the early presentations of the performance, this rather disconcerted any who enjoyed romantic dalliances *au nature* when hidden by foliage or trees because plenoptic cameras can focus on a plane behind a bush – revealing all.

6.3 PROPERTIES OF THE FOURIER TRANSFORM

6.3.1 Superposition and fingerprint analysis

Latent *fingerprints* are often of degraded quality, which can impede the recognition process. One example of using the Fourier transform in images is to restore fingerprint quality. This can use all the properties of the Fourier transform, especially superposition. Figure 6.6(a) shows a fingerprint (in blood) that is obscured by the cloth it was found on. The cloth can be removed using the superposition property of the Fourier transform: we can suppress the cloth components of the transform in Figure 6.6(c) and invert it to give a fingerprint without the cloth in Figure 6.6(d). Ideally, we would use a transform of the cloth and then subtract it, but there were no volunteers for this section of the book. The cloth is largely suppressed, though the operation could be improved. There are many approaches to fingerprint enhancement [Alonso-Fernandez et al., 2007], noting that we cover image quality improvement later in Section 6.6. One famous study aimed specifically to provide 'a fast fingerprint enhancement algorithm, which can adaptively improve the clarity of ridge and valley structures of input fingerprint images' [Hong et al., 1998]. Fingerprint matching is a natural target and this can include a filter bank [Jain et al., 2000] or local features (minutiae) [Maio & Maltoni, 1997].

There are applications where superposition does not apply. For example, when recognising people in natural scenes (in images taken outdoors) it is often required to separate a subject from their background to form either an image of just the subject or an image of just the background. The principle of superposition does not apply in that case because a subject might be obscured (say by a tree) and the subject interacts with the background's illumination (say by casting a shadow). This interaction between the subject and the background implies that the principle of superposition does not apply.

6.3.2 Invariance and image texture analysis

In image analysis, rather than use borders of objects it can be prudent to explore what is inside the borders. This is region-based analysis and the overall appearance of an object is known as its *texture*. Texture is actually a very nebulous concept, often attributed to human

FIGURE 6.6 Separating a fingerprint from its regular background: (a) original image of a fingerprint on cloth; (b) Fourier transform of (a); (c) filtered Fourier transform; (d) fingerprint without cloth

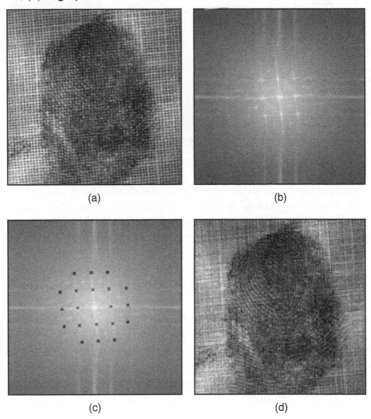

(a)   (b)

(c)   (d)

perception, as either the feel or the appearance of (woven) fabric. Everyone has their own interpretation as to the nature of texture; there is no mathematical definition for texture, it simply exists. By way of reference, let us consider one of the dictionary definitions [Thompson, 1996]:

> 'texture *n.*, & *v.t.* 1. *n.* arrangement of threads etc. in textile fabric. characteristic feel due to this; arrangement of small constituent parts, perceived structure, (*of* skin, rock, soil, organic tissue, literary work, etc.); representation of structure and detail of objects in art;...'

That covers quite a lot. If we change 'threads' for 'pixels', the definition could apply to images (except for the bit about artwork). Texture analysis has a rich history in image processing and computer vision and there are even books devoted to it [Petrou & Sevilla, 2006]. There is no

FIGURE 6.7 Three Brodatz textures: (a) French canvas (detail) D20; (b) French canvas D21; (c) Beach sand D29

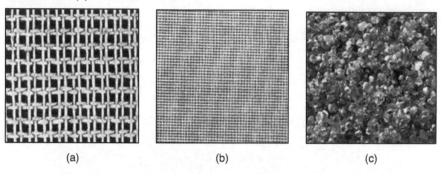

(a)   (b)   (c)

unique definition of texture and there are many ways to describe and extract it. It is a very large and exciting field of research in computer vision and there continues to be many new developments. One way to define texture is by a database of images. Three images from the early Brodatz data set of textures are shown in Figure 6.7 [Brodatz, 1966], produced originally as photographs for artists and designers, rather than for digital image analysis, but they were used a lot in early studies of image texture.

Concentrating on the word 'arrangement', we see a need for position-, scale- and rotation-invariant feature description of texture. Such a description is ubiquitous in image analysis because we seek to find and describe features wherever they might appear, however, far they were from a camera (the scale) and, however, the camera was orientated (which is image rotation). Shift invariance is a glorious tenet of the Fourier transform. Earlier, Section 4.3.4, images were jumbled up and yet still provided the same magnitude image. Given we seek to describe an arrangement of parts, it is then no surprise that early approaches to image texture analysis deployed the Fourier transform (because scale is also understood, and rotation is easily handled).

The most basic approach is to generate the FFT of an image and then to group the transform data in some way so as to obtain a set of measurements. Naturally, the size of the set of measurements is usually smaller than the size of the image's transform, so this is a different form of image coding (we are coding the texture). Here, we must remember that for display we rearrange the Fourier transform so that the d.c. component is at the centre of the presented image, though we rarely bother centring when processing image databases.

The transforms of the three Brodatz textures of Figure 6.7 are shown in Figure 6.8. Figure 6.8(a) shows a collection of frequency components, which are then replicated with the same structure (consistent with scaling the Fourier transform) in Figure 6.8(b).

FIGURE 6.8 Fourier transforms of the Brodatz textures: (a) French canvas (detail); (b) French canvas; (c) Beach sand

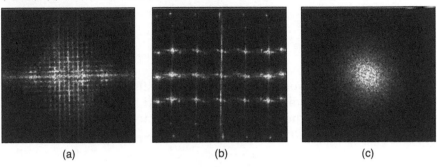

(a)   (b)   (c)

Figures 6.8(a) and (b) show the frequency scaling property of the Fourier transform: greater magnification reduces the high-frequency content. Figure 6.8(c) is clearly different in that the structure of the transform data is spread in a different manner to that of Figures 6.8(a) and (b). Naturally, these images have been derived by application of the FFT, which we shall denote as usual as

$$\mathbf{FP} = \mathcal{F}\,(\mathbf{P}) \tag{6.6}$$

where **FP** and **P** are the transform (spectral) and pixel data, respectively, and $\mathcal{F}$ is the 2-D FFT. One clear advantage of the Fourier transform is that it possesses shift invariance (Section 3.3.3): the transform of a bit of (large and uniform) cloth will be the same, whatever segment we inspect. An alternative view is that it does not matter where the camera is placed above the viewed texture, removing the need for calibration. Shift invariance is consistent with the observation that phase is of little use in Fourier-based texture systems [Pratt, 1992], so the magnitude of the transform is usually used. The transform is of the same size as the image, even though conjugate symmetry (Section 3.3.6) of the transform implies that we do not need to use all its components as measurements. As such we can filter the Fourier transform (Section 3.5) so as to select those frequency components deemed to be of interest to a particular application. Alternatively, it is convenient to collect the magnitude transform data in different ways to achieve a reduced set of measurements. First though, the transform data can be normalised by the sum of the squared values of each magnitude component (excepting the zero-frequency components, those for $u = 0$ and $v = 0$), the magnitude data is invariant to linear shifts in illumination to obtain normalised Fourier coefficients **NFP** as

$$NFP_{u,v} = \frac{|FP_{u,v}|}{\sum_{u\neq0 \wedge x\neq0} |FP_{u,v}|^2} \tag{6.7}$$

The denominator is a measure of total power as in Parseval's theorem (Section 3.3.5), so the magnitude data becomes invariant to linear shifts. The spectral data can then be described by the *entropy*, *h*, as

$$h = \sum_{u=0}^{N-1} \sum_{v=0}^{M-1} NFP_{u,v} \log \left(NFP_{u,v} \right) \tag{6.8}$$

which gives compression weighted by a measure of the information content. A uniformly distributed image would have zero entropy, so the entropy measures how much the image differs from a uniform distribution. Another measure is the *energy*, *e*, as

$$e = \sum_{u=0}^{N-1} \sum_{v=0}^{M-1} \left(NFP_{u,v} \right)^2 \tag{6.9}$$

which gives priority to larger items (by virtue of the squaring function). The measure is then appropriate when it is the larger values that are of interest; it will be of little use when the measure has a uniform distribution. Another measure is the *inertia*, *i*, defined as

$$i = \sum_{u=0}^{N-1} \sum_{v=0}^{M-1} (u - v)^2 NFP_{u,v} \tag{6.10}$$

which emphasises components that have a large separation. As such, each measure describes a different facet of the underlying data. These measures are shown for the three Brodatz textures in Table 6.1. For effective texture description/recognition, the measures should be the same for the same texture and should differ for a different one. Here, the texture measures are actually different for each of the textures. Perhaps the detail in the French canvas, Table 6.1(a), could be made to give a closer measure to that of the full resolution, Table 6.1(b), using the frequency scaling property of the Fourier transform, discussed in Section 4.3.3. The beach sand clearly gives a set of measures that are quite disparate from the other two, Table 6.1(c). In fact, the beach sand in Table 6.1(c) would appear to be more similar to the French canvas in Table 6.1(b), because the inertia and energy measures are much closer than those for Table 6.1(a) (only the entropy measure in Table 6.1(a) is closest to Table 6.1(b)). This is consistent with the images: each of the beach sand Figure 6.8(c) and French canvas Figure 6.8(b) has a large proportion of higher-frequency information, because each is a finer texture than that of the detail in the French canvas, Figure 6.8(a).

By Fourier analysis, the measures are inherently position-invariant. Clearly, the entropy, inertia and energy are relatively immune to rotation, because order is not important in their calculation. Also, the measures can be made scale-invariant, as a consequence of the scaling property of the Fourier transform. Finally, the measurements (by virtue

TABLE 6.1 Measures of the Fourier transforms of the three Brodatz textures.

(a) French canvas (detail)	(b) French canvas	(c) Beach sand
entropy(FD20) = −253.11	entropy(FD21) = −196.84	entropy(FD29) = −310.61
inertia(FD20) = 5.55 × 10^5	inertia(FD21) = 6.86 × 10^5	inertia(FD29) = 6.38 × 10^5
energy(FD20) = 5.41	energy(FD21) = 7.49	energy(FD29) = 12.37

of the normalisation process) are inherently invariant to linear changes in illumination. Naturally, the descriptions will be subject to noise. To handle large data sets we need a larger set of measurements (larger than the three given here) in order to better discriminate between different textures. There are other measures possible, naturally. Amongst others, there are elements of Liu's features [Liu & Jernigan, 1990] chosen in a way aimed to give Fourier transform-based measurements good performance in noisy conditions.

Naturally, there are many other transforms and these can confer different attributes in analysis. The wavelet transform is very popular because it allows for localisation in time and frequency [Laine & Fan, 1993]. Other approaches use the Gabor wavelet [Bovik et al., 1990; Daugman, 1993]. In fact, one survey [Randen & Husoy, 1999] includes the use of Fourier, wavelet and Discrete Cosine transforms (Chapter 5) for texture characterisation. These approaches are structural in nature: an image is viewed in terms of a transform applied to a whole image as such exposing its structure. This is like the dictionary definition of an arrangement of parts. Another part of the dictionary definition concerned detail: this can of course be exposed by analysis of the high-frequency components, but these can be prone to noise. The structural approaches lack performance compared with the very popular Local Binary Patterns (LBPs) and its variants ably summarised elsewhere [Liu et al., 2019]. These use neat tricks to describe texture, not transforms, and perform very well (as ever, transforms are used first and something else brings home the bacon). A systematic review of LBP techniques with performance evaluation is also available [Liu et al., 2017].

6.3.3   Invariance and image registration

Image registration concerns matching the positions of the contents of images and it is an enormous field in which the Fourier-based approaches are one of a selection of methods [Zitova & Flusser, 2003], as previously mentioned in Section 5.5.2. One of the earlier approaches used the invariance properties afforded by Fourier [Reddy & Chatterji, 1996]. A later approach [Pan et al., 2008] developed the Multilayer Fractional Fourier Transform (MLFFT) aimed to reduce interpolation errors in both polar (rotation- and shift-invariant matching) and log-polar (rotation-, shift- and scale-invariant) matching Fourier transforms. The study noted limitations of Fourier-based approaches as:

1. rotation aliasing;
2. scaling;
3. interpolation error; and
4. other errors, such as perspective and camera distortions.

As interpolation had previously been found to introduce error, the new approach used 'an interpolation process from a multilayer method to the real polar or log-polar grid' claiming high accuracy in registration when large-scale factors and rotations were required, with routine implementation and fast calculation. A particular criticism of previous approaches concerned the efficacy of the pseudo-log-polar transforms, which are part of the complexity associated with the Fourier–Mellin implementation. The other approaches to registration are computer vision-based and often depend on algorithms that have a lesser analytic basis than the Fourier material. It was part in response to such criticism that the OpenCV system was developed. It is not a criticism, because the application requirements of the Fourier-based approaches can lead to complicated code with many practical considerations. The reported results are certainly impressive, shown for a mosaicing example in Figure 6.9. Here, three images derived from a hand-held camera with different focus lengths in Figures 6.9(a)–(c) are registered so as to produce the mosaic in Figure 6.9(d). Note that this is before our smart cameras and is part of the work leading to the selection of image-based apps, which are now in common use. One text includes registration and many of the computer vision approaches requisite in the analysis [Cipolla et al., 2014].

6.3.4 Differentiation and image feature extraction

6.3.4.1 Template convolution and edge detection

Finding features in images by computer vision can rely on their difference in intensity from their background, as shown when detecting shapes in Section 4.4.3. After all, that is how we perceive objects by our own vision. As we have seen in Section 3.3.7, differencing is akin with differentiation. When this is applied to images, the operation is known as edge detection and the most famous operator for this is the Sobel operator. This operator consists of two templates that are convolved with the image to determine the edge data in vector form. The two differencing templates are shown in Figure 6.10. The result of their convolution with an image gives components oriented along the horizontal and vertical axes and these are combined in vector form to give magnitude and direction at the location of the centre of the template, Figure 6.10(c). The magnitude is the size of the change in brightness at that point, the direction is (or can be arranged to be) tangential, parallel to the edge. It has proved popular because it gives, overall, a better performance than other edge-detection operators of the same ilk.

The magnitude of the result of convolving the two templates with an image is shown in Figure 6.11. In order to see all the edges more clearly, the edge magnitude, Figure 6.11(b), has been thresholded to show those points that exceed a chosen value, Figure 6.11(c). The edge direction is also shown, and this is often used in computer vision algorithms. The image in Figure 6.11(c) is a form of caricaturist's sketch, though without the exaggeration that a caricaturist would imbue. It contains much of the salient image information, so it is often known as a primal sketch.

FIGURE 6.9 Registering images via a Fourier-based algorithm [Pan et al., 2008]: (a) image 1; (b) image 2; (c) image 3; (d) registration result by MLFFT

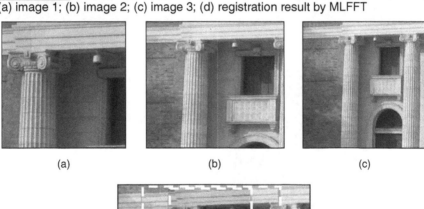

(a)   (b)   (c)

(d)

FIGURE 6.10 Templates for Sobel operator: (a) horizontal difference *Mx*; (b) vertical difference *My*; (c) vector operation

-1	0	1
-2	0	2
-1	0	1

-1	-2	-1
0	0	0
1	2	1

Gradient magnitude

Gradient direction

My

Mx

      (a)   (b)   (c)

FIGURE 6.11 Applying the Sobel operator: (a) original image; (b) edge magnitude; (c) thresholded magnitude; (d) edge direction

      (a)   (b)

      (c)   (d)

Many presentations of the Sobel operator simply give the 3×3 templates and stop there. There is an approach to generalise the operator to give bigger versions, such as 5×5 and 7×7 [Nixon & Aguado, 2019]. The templates here are smaller than the size limits specified in Section 4.4.2 for advantageous use of the FFT, but the generalised versions can be much larger.

6.3.4.2　z-transform and Fourier analysis, and edge detection operators

So how can the frequency domain help here? Let us take the discrete Fourier transform (DFT) of a 7×7 template and this is shown in Figure 6.12(a). Here, the centre axes are row and column 4, which have a peak along the y-axis and are zero along the x-axis. Looking closer, the peak at the centre of the y-axis reflects low-pass filtering and the gap at the centre of the x-axis reflects a high-pass filter. So the Sobel operator gives differentiation along one axis, the x-axis, whilst concurrently giving low-pass filtering along the other axis, the y-axis. We can see this with the Fourier transform but only infer it from the maths. In fact, this is quite a radical view of this popular operator because many texts simply give the components of Figure 6.10, without considering extension to larger templates (all you have to do is smooth the image, perhaps) and no consideration of low- and high-pass filtering, just differencing and averaging. It is worth noting that a similar analysis can be applied to the early layers of deep learning/convolutional neural net approaches to take advantage of the insight that the frequency domain provides.

Now we know the Sobel operator has a transform equivalent, we can go even further. Essentially z^{-1} is a unit time-step delay operator, so z can be thought of as a unit (time-step) advance, so $f(t - \tau) = z^{-1}f(t)$ and $f(t + \tau) = zf(t)$, where τ is the sampling interval. Given that we have two spatial axes x and y, we can then express the Sobel operator Mx of Figure 6.10(a) using delay and advance via the z-transform

FIGURE 6.12 Fourier transforms of a Sobel operator: (a) magnitude of DFT of a Sobel template from Figure 6.10(a); (b) magnitude of continuous version from Equation 6.12

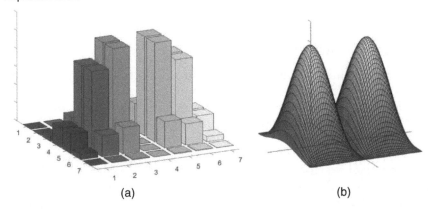

(a)　　　　　　　　　　　　　　(b)

notation along the two axes as

$$\begin{aligned} & -z_x^{-1}z_y^{-1} + 0 + z_x z_y^{-1} \\ Sobel\,(x,y) = \; & -2z_x^{-1} \quad\; + 0 + 2z_x \\ & -z_x^{-1}z_y \;\; + 0 + z_x z_y \end{aligned} \tag{6.11}$$

including zeros for the null template elements. (Advance would be a stretch in real-time signal processing; here advance can be considered because the image is stored.) Given the substitution (Section 5.6) $z^{-1} = e^{-j\omega t}$ to transform from the time (z) domain to the frequency domain (ω), then we have

$$\begin{aligned} Sobel\,(x,y) = \; & -e^{-j\omega_x t}e^{-j\omega_y t} + e^{j\omega_x t}e^{-j\omega_y t} - 2e^{-j\omega_x t} + 2e^{j\omega_x t} \\ & -e^{-j\omega_x t}e^{j\omega_y t} + e^{j\omega_x t}e^{j\omega_y t} \\ = \; & \left(-e^{-j\omega_x t} + e^{j\omega_x t}\right)\left(e^{-j\omega_y t} + 2 + e^{-j\omega_y t}\right) \end{aligned}$$

by factorisation $= \left(-e^{-j\omega_x t} + e^{j\omega_x t}\right)\left(e^{\frac{-j\omega_y t}{2}} + e^{\frac{j\omega_y t}{2}}\right)^2$

so

$$Sobel\,(x,y) = 8j\sin\left(\omega_x t\right)\cos^2\left(\frac{\omega_y t}{2}\right) \tag{6.12}$$

This gives the transform of *Sobel* as a function of spatial frequency ω_x and ω_y, along the x- and the y-axes. This confirms rather nicely the separation between smoothing along the y-axis (the cos function in the first part of Equation 6.12, low-pass) and differentiation along the other – here by differencing (the sin function, high-pass) along the x-axis. This provides the continuous form of the magnitude of the transform shown in Figure 6.12(b), which can be compared with the DFT of the template in Figure 6.12(a).

6.4   PROCESSING SIGNALS USING THE FOURIER TRANSFORM

6.4.1   Convolution theorem and ear biometrics

There are of course many image filtering operators; we have previously covered those that are amongst the most basic. There are others that offer alternative insight, sometimes developed in the context of a specific application. By way of example, Hurley developed a transform called the *force field transform* [Hurley et al., 2002; Hurley et al., 2005], which uses an analogy to gravitational force. The transform pretends that each pixel exerts a force on its neighbours, which is inversely proportional to the square of the distance between them. This generates a force field where the net force at each point is the aggregate of

FIGURE 6.13 Illustrating the force field transform [Hurley et al., 2002]: (a) image of an ear; (b) magnitude of force field transform

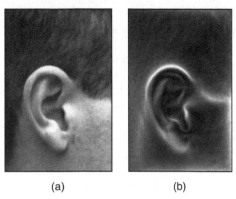

(a)   (b)

the forces exerted by all the other pixels on a 'unit test pixel' at that point. This very large-scale summation affords very powerful averaging, which reduces the effect of noise whilst retaining major features. The approach was developed in the context of ear biometrics, recognising people by their ears, which has unique advantage as a biometric in that the shape of people's ears does not change with age, and of course – unlike a face – ears do not smile! The force field transform of an ear, Figure 6.13(a), is shown in Figure 6.13(b). Here, the averaging process is reflected in the reduction of the effects of hair. The transform has also highlighted ear structures, especially the top of the ear and the lower 'keyhole' (the inter-tragic notch).

The image shown in Figure 6.13(b) is actually the magnitude of the force field. The transform itself is a vector operation and includes direction [Hurley et al., 2002]. The transform is expressed as the calculation of the force **F** between two points at positions r_i and r_j, which is dependent on the value of a pixel at point r_i as

$$F_i(r_j) = P(r_i) \frac{r_i - r_j}{|r_i - r_j|^3} \tag{6.13}$$

which assumes that the point r_j is of unit 'mass'. This is a directional force (which is why the inverse square law is expressed as the ratio of the difference to its magnitude cubed) and the magnitude and directional information has been exploited to determine an ear 'signature' by which people can be recognised. In application, Equation 6.13 can be used to define the coefficients of a template that is convolved with an image; an implementation is also given [Hurley et al., 2002] and this uses the FFT to improve speed. In fact, it is not just an improvement as we found it absolutely necessary. Without the FFT it is a go-and-make-and-then-eat-dinner algorithm (and do the washing up), whereas it is virtually real-time with the FFT. Many have found the advantage of using the FFT: for example, it would be impossible to study and deploy wavelets without it.

6.4.2 Deconvolution and image enhancement

When discussing convolution in previous chapters, the prospect of the inverse process, *deconvolution*, has not been raised. This is because practical problems make deconvolution suited to fixed applications and it is sometimes considered to be largely of theoretical interest. It did offer a famous solution in one case and is used in specialised cases. Conceptually, deconvolution is quite simple and that makes it attractive. The output of a system $o(t)$ given by convolution is

$$o(t) = p(t) * q(t) \tag{6.14}$$

where the system function is $p(t)$ and the function $q(t)$ is some function that degrades the output, say a muffler on a microphone or a lack of focus in a 2-D image. Naturally, we would seek to remove the degradation, and the nature of convolution in the time domain makes it unclear how this can be achieved. By the convolution theorem,

$$o(\omega) = p(\omega) \times q(\omega) \tag{6.15}$$

So to remove the degradation, we simply need to divide the output by the function $q(\omega)$

$$p(t) = \mathcal{F}^{-1}\left(\frac{o(\omega)}{q(\omega)}\right) \tag{6.16}$$

and hey presto, we have a decontaminated output. Naturally, it is more complicated than this procedure and we shall consider that soon. Deconvolution was famously used in the Hubble telescope when poor-quality control in production led to a lens aberration, which contaminated the images of space. This was rather embarrassing (and costly) because the telescope had already been launched, with much fanfare. The aberration was corrected by deconvolution when a correction function was estimated to correct the images, though the deconvolution function appears less regular than expected. This was presumably not of consistent quality, because the telescope was later fixed by optical means, sacrificing other equipment. Things appear to get even better when we use an optimisation process to remove an unknown function, using *blind deconvolution* [Campisi & Egiazarian, 2017; Kundur & Hatzinakos, 1996]. A single image restored using blind deconvolution is shown in Figure 6.14(b), which shows considerably greater detail than the original frame, Figure 6.14(a).

Alles ist gut until up pops life and chucks a spanner into the works: if the signal is contaminated by noise, or $q(\omega)$ is not known precisely,

FIGURE 6.14 Blind deconvolution applied to Hubble imagery (with permission from [Schulz et al., 1997] © The Optical Society): (a) original frame; (b) frame restored by blind deconvolution

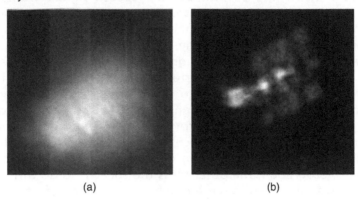

(a)   (b)

then the reconstruction will not be correct. Noise is not just a feature of UK politics: it is anywhere and everywhere and random values cannot be predicted. One of our wheels has clearly fallen off, especially so when $q(\omega)$ might sometimes be zero and so division fails (for two dimensions the process uses matrix inversion thus requiring a square image, which needs also to be non-singular). Solutions have been shown to be possible, with counsel of the form that an ill-posed problem can be turned into a well-posed one by the addition of a regularising function, but it is rather hard to avoid the central problems with the process (some major textbooks in this field do not even mention deconvolution [e.g. Oppenheim et al., 1997]). Should you seek guidance, one place to start is *Tikhonov regularisation*. Clearly, deconvolution is of use and has been used in fixed applications like the Hubble telescope and, say, deblurring in confocal microscopy. Clearly, deconvolution also has problems and in cases where noise is to be reduced, or the contaminant function cannot be estimated, there are other popular means of improving quality as we shall find in Section 6.6, or it can be sharpened by unsharp masking, Section 4.4.4.2.

6.4.3 Speech recognition and correlation

Pattern recognition and artificial intelligence apply much in our society now. For a foray into recognition, we shall consider 1-D signals: *speech recognition*. There are other applications of recognition by 1-D signals such as *electroencephalogram* (EEG) and *electrocardiogram* (ECG) biometrics for person recognition, and some of the techniques are very similar. Recognition requires matching, and we considered the correlation of 1-D signals previously in Section 3.4.2.

Speech recognition has come on a long way in recent years, with Alexa and Siri being used to enrich and ease our daily lives. To do this, the systems have to pick out the speech signals from a fog of noise: speech is recognised even when the kettle's boiling and the dog is barking (OK, a bit of a stretch there). But it is actually fantastic that it can be done at all. So how is it done? One can match time-domain speech signals using dynamic time warping which avoids the problem of correlation with time-variant signals, but that rather lacks insight into the process, and that insight is needed to devolve the speech signal into a selection of its components, which can be used to improve the matching process. The first point in the process is that speech is a function of the vocal tract, and the speech signal itself. If we recognise the vocal tract, then it is *speaker recognition*, the other bit is the speech. To do this we can use *cepstral* analysis, which is essentially the transform of a transform. The words used are classic academic gamesmanship using words from the frequency domain in a form of an anagram. Spectrum becomes *cepstrum, frequency* becomes *quefrency*, similarly we find *alanysis, liftering* (agh!) and *saphe* (oh, the importance of saphe eh!). The process is shown in Figure 6.15.

This terminology is pretty horrible; it's a load of loblocks!

Because a cepstrum is the transform of a transform, it measures the frequency of frequency (that should perhaps be expressed as quefrency of frequency). It is used in speech and speaker recognition because the effects of the speech (the pitch) and the vocal tract (the formants) are additive in the quefrency domain and are separated in the logarithm of the power spectrum. This allows for separation of the signals, which is a form of deconvolution allowing investigation of periodic structures within unknown spectra. In this way they can be used to analyse speech, radar and EEG/ ECG signals. Mathematically, we have that speech sound (t) is the response of the pitch $s(t)$ and the vocal tract $v(t)$

$$\text{sound}(t) = s(t) * v(t) \tag{6.17}$$

by the convolution theorem, using the Short-Time Fourier transform (Section 2.5.1) of a speech segment

$$\text{sound}(\omega) = \mathcal{F}(s(t) * v(t)) = Fs(\omega) \times Fv(\omega) \tag{6.18}$$

FIGURE 6.15 Cepstral analysis in speech recognition

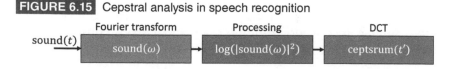

and from the magnitude (discarding the saphe, sic)

$$\log\left(|\text{sound}\,(\omega)|\right) = \log\left(|Fs\,(\omega)|\right) + \log\left(|Fv\,(\omega)|\right) \qquad (6.19)$$

In this way, the sound signal is decomposed into its cepstral components of speech $cs\,(t)$ and voice $cv\,(t)$ using the DCT as

$$\text{unsod}\,(t') = \left|\text{DCT}\left(\log\left(|Fs\,(\omega)|^2\right) + \log\left(|Fv\,(\omega)|^2\right)\right)\right|^2$$
$$= c\,(t') = cs\,(t') + cv\,(t') \qquad (6.20)$$

The process is illustrated in Figure 6.16, for recorded speech. Here samples of the word 'Fourier' uttered by a single speaker (guess who), Figure 6.16(a), are shown as a Fourier transform (Equation 6.19) in Figure 6.16(b), and then as its cepstrum (Equation 6.20) in Figure 6.16(c). The remarkable energy compaction of the process is revealed in the details of the Fourier transform and of the cepstrum, Figures 6.16(d) and (e), which leads to significantly fewer components in the cepstrum. It is after all, the transform of a transform. The cepstrum here includes both the speaker and the speech as in Equation 6.20. To determine which is which we would need to design a data set comprising recordings of different words from different speakers under controlled recording conditions.

The use of magnitude in the process is a general assumption that phase is of little consequence when analysing longitudinal waves (we are not interested in reverberations), though there are studies that suggest phase is important to speech intelligibility for humans (e.g. Alsteris & Paliwal, 2007). Because human hearing appears to be most discriminative at low frequencies and least at high frequencies a logarithmic *mel-frequency* scale is used rather than a pure logarithm aiming to relate the perceived frequency of a tone to its actual frequency. Then *Mel-Frequency Cepstral Coefficients* (*MFCCs*) are used to describe the speech signal, using a DCT to separate the frequency components. Now we have extracted the different features, we need to perform recognition.

There are many ways to achieve recognition and amongst the most popular are *Hidden Markov Models* (*HMMs*) which essentially model speech as a chain of events, and these are classified using *Support Vector Machines* (*SVMs*). HMMs have dominated speech recognition for some time now [Juang & Rabiner, 1991]. There is the KALDI open-source software [Povey et al., 2011] which has MFCC and perceptual linear prediction (which uses a cube root rather than a log function for amplitude compression) for feature extraction and HMMs and GMMs for acoustical modelling. This gives for optimisation of many parameters, especially when parts are included, which have not been

FIGURE 6.16 Cepstrum of sound: (a) sound; (b) spectrum, $\log\left(|Fsound_u|\right)$; (c) cepstrum, $unsod_{t'}$; (d) spectrum – detail; (e) cepstrum – detail

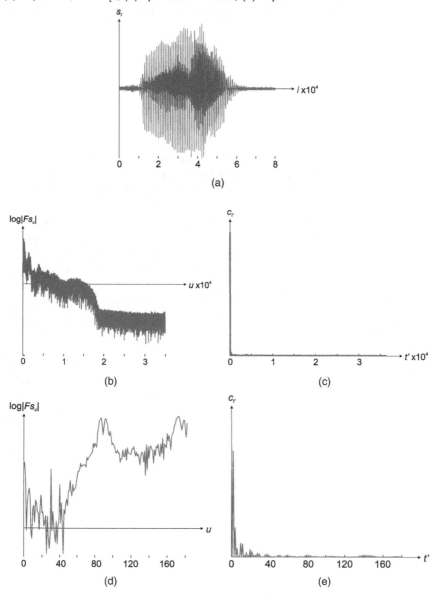

described here, such as the window functions. Nowadays the interest has shifted to deep learning [Graves et al., 2013] and end-to-end systems for recognition, given the improved performance that can be achieved. Speech recognition systems will just get better, and we can indeed switch the kettle on.

This section has not mentioned correlation a great deal. Correlation is rarely to be found as a major topic within applications, because it is usually a part of the toolset used. In image analysis, correlation becomes template/image matching as described earlier. A popular way used to find shapes in images is called the Hough transform (it is not a transform like the Fourier transform, it is a mapping from the image to evidence that a shape exists). It has been shown that there is a direct link between the Hough transform and template matching [Aguado et al., 2000] and so by implication correlation is often used to find shapes in computer vision, though it is rarely described as such. Template matching can use the Fourier transform for reasons of computation, as described earlier. Many deep learning architectures are called convolutional neural networks, but they actually use correlation because the convolution functions are even.

6.5   THE IMPORTANCE OF PHASE AND PHASE CONGRUENCY

We have already seen the importance of phase in beamforming: without phase, we could not estimate direction. Here we shall see the notion of phase underpinning an image feature extractor, and one with impressive performance indeed. Its complexity means that it is rarely seen in image-processing packages and textbooks, which is a pity.

Edge detection operators in image processing and computer vision have innate problems: their application can derive incomplete contours, invariably there is a need for selective thresholding; and because their basis is differentiation there are often problems with noise. Changes in local illumination exacerbate difficulty in threshold selection. In computer vision, some of these problems can be handled within shape extraction, which is at a higher level of operation, when the extraction process can be arranged to accommodate partial data and to reject spurious information. Naturally, there is interest in refining the low-level feature extraction techniques further and there have been many studies on edge detection. The Fourier transform provides an alternative basis for their analysis, as we have just seen.

Phase congruency is a feature detector with two main advantages: it can detect a broad range of features; and it is invariant to local (and smooth) change in illumination [(Kovesi, 1999]. As the name suggests, it is derived by frequency-domain considerations operating on the considerations of phase (a.k.a. time). It is illustrated detecting some 1-D features in Figure 6.17, where the data are the points ∗ and contain: a (noisy) step function in Figure 6.17(a); and a peak (or pulse) in Figure 6.17(b). By Fourier transform analysis, any function is made up from the controlled addition of sine waves of differing frequencies

FIGURE 6.17 Low-level feature extraction by phase congruency: (a) step edge; (b) peak

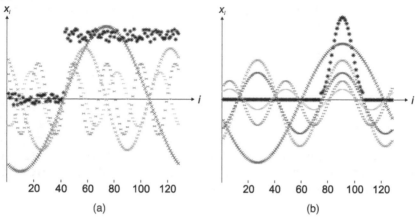

(a)   (b)

(here shown with magnitude chosen for illustration only). For the step to occur, constituent frequencies – the points X in Figure 6.17(a) – must have an edge or a zero-crossing, going from negative to positive, at the same time. In this way they add up to give the edge because they are all congruent at the edge and cancel out when the signal is zero. Similarly, for the peak to occur, then the constituent frequencies must have a congruent peak at the same time; in Figure 6.17(b) the points ∗ are the peak (actually a Hamming window) and the points X are from some of its constituent frequencies. This means that to find a feature we are interested in, we can determine points where events happen at the same time: this is phase congruency. By way of generalisation, a triangle wave is made of peaks and troughs: phase congruency implies that the peaks and troughs of the constituent signals should coincide.

In fact, the constituent sine waves plotted in Figure 6.17(a) were derived by taking the Fourier transform to determine the sine waves according to their magnitude and phase. The DFT in Equation 3.6 delivers the complex Fourier components. These can be used to show the constituent signals p_x by

$$p_x = |Fp_u| e^{j\left(\frac{2\pi}{N} ux - \varnothing(Fp_u)\right)}$$
(6.21)

where $|Fp_u|$ is again the magnitude of the uth Fourier component and $\varnothing(Fp_u)$ is the argument, the phase. The frequencies (x) displayed in Figure 6.17 are the first components. The addition of these components is indeed the inverse Fourier transform which reconstructs the step feature.

FIGURE 6.18 Edge detection by the Canny operator and by phase congruency: (a) modified cameraman image; (b) edges by the Canny operator; (c) phase congruency

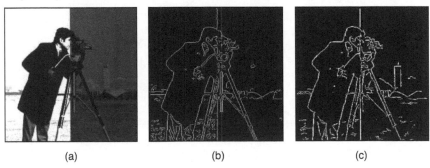

(a)   (b)   (c)

The advantages are that detection of congruency is invariant with local contrast: the sine waves still add up so the changes are congruent, even if the magnitude of the step edge is much smaller. In images, this implies that we can change the contrast and still detect edges. This is illustrated in Figure 6.18. Here a standard image, the 'cameraman' image from the early UCSD data set has been changed between the left and right sides so the contrast changes in the two halves of the image, Figure 6.18(a). Edges detected by the popular *Canny* operator (a popular and more sophisticated operator than Sobel) are shown in Figure 6.18(b) and by phase congruency in 6.18(c). The basic structure of the edges detected by phase congruency is very similar to that structure detected by Canny, and the phase congruency edges appear somewhat cleaner (there is a single line associated with the tripod control in phase congruency); both detect the change in brightness between the two halves. There is a major difference though: the building in the lower right side of the image is barely detected in the Canny image whereas it can clearly be seen by phase congruency. Its absence is because of the parameter settings used in the Canny operator. These can be changed, but if the contrast were to change again, then the parameters would need to be re-optimised for the new arrangement. This is not the case for phase congruency.

The original notions of phase congruency were the concepts of *local energy* [Morrone & Owens, 1987], with links to the human visual system. One of the most sophisticated implementations was by Kovesi [Kovesi, 1999], with added advantage that his Matlab implementation is available on the Web (https://www.peterkovesi.com/matlabfns/) as well as much more information. Essentially, we seek to determine features by detection of points at which Fourier components are maximally in phase. By extension of the Fourier reconstruction functions in Equation 6.21, Morrone and Owens defined a measure of phase

FIGURE 6.19 Summation in phase congruency

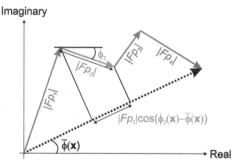

congruency *PC* as

$$PC(\mathbf{x}) = \max_{\overline{\varnothing}(\mathbf{x}) \in 0, 2\pi} \left(\frac{\sum_u |Fp_u| \cos\left(\varnothing_u(\mathbf{x}) - \overline{\varnothing}_u(\mathbf{x})\right)}{\sum_u |Fp_u|} \right) \qquad (6.22)$$

where $\varnothing_u(\mathbf{x})$ represents the local phase of the component Fp_u. Essentially this computes the ratio of the sum of projections onto a vector (the sum in the numerator) to the total vector length (the sum in the denominator). The value of $\overline{\varnothing}_u(\mathbf{x})$ that maximises this equation is the amplitude weighted mean local phase angle of all the Fourier terms at the point being considered. In Figure 6.19 the resulting vector is made up of four components, highlighting the projection of the second onto the resulting vector. Clearly, the value of *PC* varies between 0 and 1, and the maximum occurs when all elements point along the vector. As such, the resulting phase congruency is a dimensionless normalised measure, which can be thresholded for image analysis.

In this way, we have calculated the phase congruency for the noisy step function in Figure 6.20(a), which is shown in Figure 6.20(b). Here, the position of the step is at time step 40; this is the position of the peak in phase congruency, as required. Note that the noise can be seen to affect the result, though the phase congruency is largest at the right place.

One interpretation of the measure is that because for small angles, $\cos\theta = 1 - \theta^2$, Equation 6.22 expresses the ratio of the magnitudes weighted by the variance of the difference to the summed magnitude of the components. There is difficulty with this measure, apart from the difficulty in implementation: it is sensitive to noise, as is any phase measure; it is not conditioned by the magnitude of a response (small responses are not discounted); and it is not well localised (the measure varies with the cosine of the difference in phase, not with the difference itself – though it does avoid discontinuity problems with direct

FIGURE 6.20 One-dimensional phase congruency: (a) (noisy) step function; (b) phase congruency of (noisy) step function

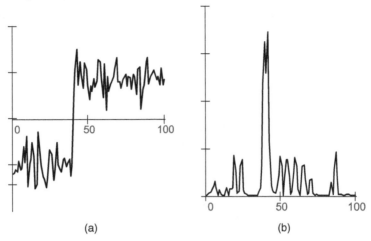

(a)   (b)

use of angles). In fact, the phase congruency is directly proportional to the local energy, so an alternative approach is to search for maxima in the local energy. The notion of local energy allows us to compensate for the sensitivity to the detection of phase in noisy situations.

For these reasons, Kovesi developed a wavelet-based measure, which improved performance, whilst accommodating noise. In basic form, phase congruency can be determined by convolving a set of wavelet filters with an image and calculating the difference between the average filter response and the individual filter responses. The response of a (1-D) signal I to a set of wavelets at scale n is derived from the convolution of the cosine and sine wavelets (discussed in Section 5.7.2) denoted M_n^e and M_n^o, respectively

$$\left[e_n(\mathbf{x}), o_n(\mathbf{x})\right] = \left[I(\mathbf{x}) * M_n^e, I(\mathbf{x}) * M_n^o\right] \qquad (6.23)$$

to deliver the even and odd components at the nth scale $e_n(\mathbf{x})$ and $o_n(\mathbf{x})$, respectively. The amplitude of the transform result at this scale is the local energy

$$A_n(\mathbf{x}) = \sqrt{e_n(\mathbf{x})^2 + o_n(\mathbf{x})^2} \qquad (6.24)$$

At each point $\mathbf{x}$, we will have an array of vectors that correspond to each scale of the filter. Given that we are only interested in phase congruency that occurs over a wide range of frequencies (rather than just at a couple of scales), the set of wavelet filters needs to be designed so that adjacent components overlap. By summing the even and odd components, we obtain

$$F(\mathbf{x}) = \sum_n e_n(\mathbf{x})$$

$$H(\mathbf{x}) = \sum_n o_n(\mathbf{x}) \tag{6.25}$$

and a measure of the total energy A as

$$\sum_n A_n(\mathbf{x}) \cong \sum_n \sqrt{e_n(\mathbf{x})^2 + o_n(\mathbf{x})^2} \tag{6.26}$$

then a measure of phase congruency is

$$PC(\mathbf{x}) = \frac{\sqrt{F(\mathbf{x})^2 + H(\mathbf{x})^2}}{\sum_n A_n(\mathbf{x}) + \varepsilon} \tag{6.27}$$

where the addition of a small factor ε in the denominator avoids division by zero and any potential result when values of the numerator are very small. This gives a measure of phase congruency, which is essentially a measure of the local energy. Kovesi improved on this, improving on the response to noise, developing a measure which reflects the confidence that the signal is significant relative to the noise. Further, he considered in detail the frequency-domain considerations, and its extension to two dimensions [Kovesi, 1999]. For 2-D (image) analysis, given that phase congruency can be determined by convolving a set of wavelet filters with an image, and calculating the difference between the average filter response and the individual filter responses. The filters are constructed in the frequency domain using complementary spreading functions; the filters were constructed in the Fourier domain because the log-Gabor function has a singularity at zero frequency

The discriminatory power of phase congruency is shown in Figure 6.21 (from Kovesi's original paper). Phase congruency picks out the windows of the houses very clearly whilst they are unclear in the Canny data. As ever it is a tradeoff between complexity and time, but with the storage and processing power of modern computers, why not plump for the power of phase congruency?

Naturally, the change in brightness in Figure 6.18 might appear unlikely in practical applications, but this is not the case with moving objects that interact with illumination or in fixed applications where the illumination can change. In studies aimed to extract spinal information from digital videofluoroscopic X-ray images in order to provide guidance for surgeons [Zheng et al., 2004], phase congruency was found to be immune to the changes in contrast caused by slippage of the lead shield used to protect the patient whilst acquiring the image information. One such image is shown in Figure 6.22. The lack of shielding is apparent in the bloom at the side of the images (the lead prevents irradiation so when it slips the image brightens). This changes

FIGURE 6.21 Edge detection by the Canny operator and by phase congruency [Kovesi, 1999]: (a) steep street; (b) edges by Canny; (c) edges by phase congruency

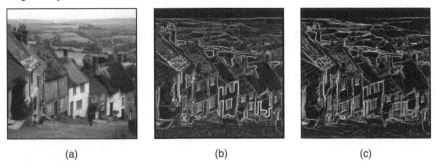

(a)   (b)   (c)

FIGURE 6.22 Vertebrae by phase congruency [Zheng et al., 2004]: (a) digital videofluoroscopic image of lower spine showing vertebrae; (b) edges by the Canny operator; (c) features by phase congruency

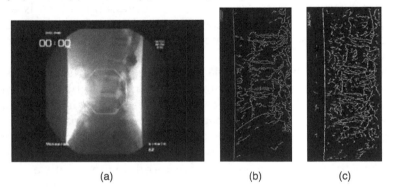

(a)   (b)   (c)

as the subject is moved so it proved difficult to optimise the parameters for a standard edge detector over the whole sequence, the Canny operator Figure 6.22(b), but the detail of a section of the phase congruency result, Figure 6.22(c), shows that when phase congruency is used for edge detection, the vertebrae information is readily available for later high-level feature extraction.

Keywords re-occur within phase congruency: Fourier, frequency domain, wavelets and convolution. The performance of phase congruency certainly encourages its consideration, especially if local illumination is likely to vary and if a range of features is to be considered. It is derived by an alternative conceptual basis, and this gives different insight, let alone performance. Even better, there is a Matlab implementation available, for application to images – allowing you to replicate its excellent results. There has been further research, noting especially an advanced operator with superior noise suppression [Xiang et al., 2017], its extension in ultrasound image analysis [Mulet-Parada & Noble, 2000] and its extension to spatiotemporal form [Myerscough & Nixon, 2004].

6.6 FILTERING AND DENOISING, AND IMAGE ENHANCEMENT

There are many applications that require improvement to image quality. Space and astronomical images are often of poor quality and might even need rectification of basic sensing errors as earlier in the Hubble telescope. Medical imaging allows us to peer inside people's bodies, which was a largely unheralded revolution of the 20th Century. The technologies include MRI, ultrasound and X rays, and though there is constant interest in improving sensor quality, there is natural interest in post hoc image quality improvement. For terminology, we have *enhancement*, which aims to improve the visual appearance. We also have *restoration*, which is to return to the original image without the corrupting artefacts. Finally, there is *denoising*, which is aimed to remove one of the many types of noise that can corrupt images. These terms are not actually synonymous but are sometimes used in that way.

The Fourier transform can play a part in this, by its access to frequency components and for computation, and it can be used to filter images (Section 4.4.4) for enhancement but its properties are ill-suited to restoration and denoising because some of the particular noise sources found in images, such as *speckle* noise in SAR and ultrasound images, are not readily amenable to a frequency-domain interpretation. One study used the 'sequence ordered' orthogonal transforms, including the Fourier, Hartley, cosine and Hadamard transforms, for image enhancement for detection and visualisation on objects within an image [Agaian et al., 2001].

In computer vision/image processing, there is a selection of state-of-art operators, which can use the Fourier transform in implementation rather than as a basis. The general approach is to preserve feature boundaries (so that objects can be detected), whilst reducing variation and noise. One of the most popular operators is the *non-local means* operator [Buades et al., 2005], which is an extended version of an averaging operator. The basic premise of the operator is to assign a point a value that is the mean of an area that is closest to the mean at the value of the point, rather than the mean at that point. As the original paper put it 'the denoised value at x is a mean of the values of all points whose Gaussian neighbourhood looks like the neighbourhood of x'. Essentially it concerns an area p and an area q, which are described by their mean values $\bar{x}(p, N)$ and $\bar{x}(q, N)$, respectively, both calculated over an $N \times N$ region. A Gaussian-weighted difference between the

FIGURE 6.23 Reducing image noise by non-local means and Gaussian operators: (a) original image; (b) Gaussian averaging; (c) non-local means

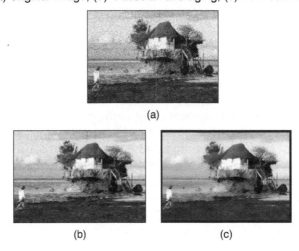

(a)

(b)   (c)

means $g\left(\overline{x}(p) - \overline{x}(q), \sigma\right)$ is minimum when they are the same and maximum when they are greatly different, so by applying an $M \times M$ search region, we have the new point, which is calculated by a weighted summation

$$N_{x,y} = \frac{1}{\sum_{i \in M} \sum_{j \in M} g\left(\overline{x}(p) - \overline{x}(q)\right)} \sum_{i \in M} \sum_{j \in M} P_{x(i),y(j)} \left(g\left(\overline{x}(p,N) - \overline{x}(q,N), \sigma\right)\right)$$

(6.28)

The parameters that must be chosen are the window size of the averaging operator, N, the size of the search region, M, and the standard deviation σ. By Figure 6.23, it can provide very impressive results, and by its simplicity and performance, it is thus a very popular operator. Other popular operators include *anisotropic diffusion* and *bilateral filtering*. So this is not actually an example of an application of the Fourier transform because Fourier gives basic filtering, not advanced processing. Figure 6.23 is more an example of what can be achieved by other means (agh, a contender for worst pun in the book there). Many techniques have been used within image restoration and denoising such as wavelets, statistics and maximum entropy [Katsaggelos, 2012] and modern approaches use deep learning: there will ever been an interest in improving image quality.

6.7   VARIANTS OF THE FOURIER TRANSFORM, AND CODING

Space costs money, so there is a continuing need for efficient use of storage, especially in images. Likewise, spectrum and power cost money in communications, so there is a continuing need for efficiency there too. As much storage as we have, we tend to need more. So there is a massive interest in *coding*. In communications, source coding refers to compression of speech and images, whereas channel coding is the communication thereof (which provides error correction). Luckily for us, some researchers have been involved in the *Joint Photographic Experts Group* (*JPEG*), and that has led to a worldwide open-source (free and widely available) standard for imagery. Clearly it has been around for a while, by the word 'photograph'. Commerce is part of life and images are now central to life (and one cannot patent a wheel thank goodness, notwithstanding some rather frustrating patents) but as we shall find there are continuing business (self-) interests, which can get in the way. Having aired that point, we shall now consider some of the major coding approaches, where we shall find that the energy compaction by transforms can be used to good effect.

> I remember someone once saying 'imagine what you can do with 64kB of memory', it was a long time ago but it seems laughable now. We will ever need more storage space, or better coding.

A major difference in coding schemes is lossy and lossless coding. Because effects of coding can be imperceptible to human vision, it can be acceptable to lose some parts of an image in the coding process without visible concern or effect – this is *lossy coding*. On the other hand, some images are used for measurement, and so the original image should be retained, even if its form is compressed (one can reconstruct the original image from its coded version) – this is *lossless coding*. Apart from image compression, many use Winzip for file compression: it uses lossless coding, obviously (and does not use Fourier). Video can generate an enormous amount of data and many approaches use lossy coding for that reason. There are many acronyms in coding, and the main approaches are summarised in Table 6.2. Another aspect is that the coding can be focussed on the material being communicated: one uses teleconferencing for speech rather than music and musical instruments can sound terrible when played over a phone line.

> In the Covid outbreak some bands tried to use teleconferencing for practice sessions, and it required some technical modifications to achieve an acceptable quality. Teleconferencing's designed for speech, not music and instruments.

There is a large difference between speech and image compression, apparently by virtue of process. Speech is used to transmit information live, whereas images are used to convey it in stored form. In some cases

TABLE 6.2	Some of the major coding schemes in current use.		
Technique	**Lossy?**	**Target**	**Basis**
Winzip	Lossless	Files	Lempel–Ziv–Storer–Szymanski/ Huffman
LPC (Linear Predictive Coding)	Lossy	Speech	LPC
MDCT (Modified Discrete Cosine Transform)	Lossy	Speech	Transform (DCT)
WMA (Windows Media Audio)	Lossy & Lossless	Speech	Transform (DCT)
GIF (Graphics Interchange Format)	Lossless	Images	Lempel–Ziv–Welch
PNG (Portable Network Graphics)	Lossless	Images	Lempel–Ziv/ Huffman
JPEG (Joint Photographic Experts Group)	Lossy	Images	Transform (DCT)
JPEG 2000	Lossy & Lossless	Images	Transform (Wavelets)
MPEG (Moving Picture Experts Group)	Lossy	Video	Transform (DCT)

speech information must be conveyed securely and images need to be encrypted. The two main approaches to each use either a transform-based approach, or an alternative non transform-based one. Noting that we have previously considered speech recognition (Section 6.4.3), the main approaches in speech use *Linear Predictive Coding (LPC*, e.g. used in telephones and MPEG-4 and FLAC audio codecs) or *MDCT* (Modified DCT used in voice over internet such as Apple's FaceTime). Lossless WMA has been used for the 1-D real signals herein.

In image coding, the main difference in technique is between transform and non-transform-based coding, with *PNG* based on the Lempel–Ziv algorithm and JPEG using the DCT. We shall not consider here the further process that includes bit allocation and quantisation aimed to minimise distortion of the signal restored from the transform. The JPEG standard includes a lossless coding mode, but this is rarely used. The newer standard *JPEG 2000* is not used as widely as expected, though it is used in film technology. The JPEG coding scheme is shown in Figure 6.24 with the DCT at the beginning. In Section 5.4 we found the DCT to result in the most compressed form of a single image that has been found in practice and why it is at the heart of this popular image-coding approach. JPEG 2000 uses wavelets, taking advantage of their sensitivity to space and frequency in an efficient coding process. This text has used lossless PNG throughout, except in Figure 6.25 (the

FIGURE 6.24 JPEG coding

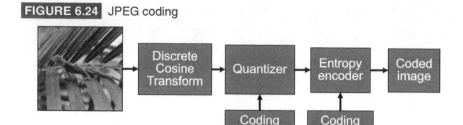

FIGURE 6.25 Frequency-domain effects of lossy and lossless coding:
(a) frequency components of lossy JPEG image, coded at low quality;
(b) frequency components of lossless PNG image

(a)   (b)

RAW format has a rather high storage cost). Coding video imposes the most exacting requirements as video is much a feature of our lives and it requires much storage, necessitating lossy coding. *MPEG* is a lossy approach using motion-compensated DCT though there is much commercial interest, which has led to the vast range of codecs that are currently used. Users may wish for greater standardisation in video coding as is available in image coding, but the commercial considerations currently prevent it. Perhaps someday when 3-D television is in general use, a ubiquitous standard will be developed.

It is possible to vary the *quality* in JPEG images. With lower quality, the image can be stored in reduced form. The effects of quality are often seen in artefacts around edges in the image. This is clearly shown in the transform, Figure 6.25(a), where low quality has introduced high-frequency data in the transform, when compared with the transform of PNG coding (or a high-quality JPEG image), Figure 6.25(b). The effect of low quality for this image has reduced the storage considerably, from 25 to 1 KB. A succession of JPEG codings can reduce quality enormously, which does not happen with lossless coding.

This short introduction to coding shows that transform-based approaches have a large role to play in coding systems. There is a vast amount of work in this field, and their achievements are excellent. There are many patents and papers as the research continues to evolve. There is modern guidance on this field [Bull, 2014; Ngan et al., 2018], but this field moves fast.

6.8 SUMMARY

This chapter has described a selection of applications of the Fourier transform, either in direct physical application, or in technique, or for improving understanding. That is what the Fourier transform is for. There are applications that have not been described and their omission is by space, because the applications herein have been chosen to illuminate all the previous material in the book. The selection of applications chosen also follows the original presentation of the continuous Fourier transform, in Chapter 2, which seems apposite though continuous and discrete are intermingled here. This chapter then follows the later structure of the book as precisely as possible. We use transforms continually in our daily lives, in technology and even in our own hearing as in Chapter 1. Arguably, the variants of the Fourier transform are themselves applications of it because without its motivation or insight they might not have been developed at all. That might seem obscure, but it does rather depend on the origin of the Fourier transform itself, as we shall find in the next chapter.

CHAPTER 7

Who and What was Fourier?

7.1 NATURE AND ORIGINS OF THE FOURIER TRANSFORM

This chapter describes the nature and origins of the Fourier transform and allows us to clear up some inconsistencies in the presentation of the Fourier transform. These have only been mentioned previously and become apparent from the study of much of the material that is available concerning the Fourier transform. We shall also consider its history, both in Fourier's work and in context, before a short consideration of Fourier and his life.

7.1.1 The basic nature and definitions of the Fourier transform

Some of the inconsistencies in the Fourier transform arise from the lack of an original definition. This is not a criticism because Fourier was sailing in uncharted waters, and in ones of incredible consequence. There is an excellent account of the history [Dominguez, 2016] and unlike that we shall consider just the DFT because it is perhaps the most used of the Fourier transform equations (generally via the FFT). In Equation 3.6, the DFT was defined as

$$Fp_u = \frac{1}{N} \sum_{x=0}^{N-1} p_x e^{-j\frac{2\pi}{N}xu} \qquad u = 0, 1 \dots N-1 \qquad (7.1)$$

with a scaling coefficient $1/N$, which was introduced to make the zero-order term equal to the average of the sampled signal **p**. This scaling gives meaning to the first-order component, which is one of the most informative in the transform data, and by the previous analysis the other components equate to realistic values. The scaling is, however, often omitted from the forward transform, and a more common definition of the DFT calculates components **Fp**$'$ using

$$Fp'_u = \sum_{x=0}^{N-1} p_x e^{-j\frac{2\pi}{N}xu} \qquad (7.2)$$

with the inverse DFT requiring a scaling coefficient as

$$p_x = \frac{1}{N} \sum_{u=0}^{N-1} Fp'_u e^{j\frac{2\pi}{N}xu} \qquad (7.3)$$

Because the inverse DFT is used only when reconstruction is required, this saves computation, however slight. It also loses natural interpretation of the coefficients because for signals with many samples the numbers become large. When reconstruction is implicit, say in filtering and convolution, there is no computational advantage in scaling (in convolution, when one of the transforms is pre-computed). Key point 10 summarises the points on generality.

Key point 10

There is no universally accepted unique definition of the Fourier transform.

There is, however, a more glaring inconsistency: why use $e^{-j\theta} = \cos\theta - j\sin\theta$ when it is possible instead to use the positive form, $e^{j\theta}$? Using the positive form gives an alternative form of the Fourier transform that can be used to calculate and deploy components **aFp** via the alternative Fourier pair

$$aFp_u = \sum_{x=0}^{N-1} p_x e^{j\frac{2\pi}{N}xu} \qquad p_x = \frac{1}{N} \sum_{u=0}^{N-1} aFp_u e^{-j\frac{2\pi}{N}xu} \qquad (7.4)$$

This appears to be used, for example, in probability theory https://en.wikipedia.org/wiki/Characteristic_function_(probability_theory) (which notes that the difference from the more usual form of the Fourier transform 'is essentially a change of parameter'). As this book is an applied rather than a theoretical study, it is clearly possible to perform the transform using Equation 7.4, though its properties might be less conventional. Figure 7.1 shows the application of Equation 7.4 to the chord of Figure 3.13 giving a magnitude spectrum, Figure 7.1(a), the difference from the conventional DFT in magnitude, Figure 7.1(b), and the difference from the conventional DFT in phase, Figure 7.1(c). The difference is in the phase rather than in the magnitude (as expected). Naturally, the inverse transform returns to the original signal, but that is not worth showing here.

Part of the dichotomy in this analysis is because of the lack of an original definition, and it may also be because of the custom and the evolution of computing. The original suggested by Fourier [Fourier, 1878] is not the form we know it now and he stated 'This equation serves to develop an arbitrary function $f(x)$ in a series of sines or cosines of multiple arcs' and he described a continuous transform (Babbage was working on the difference engine at the time, and the first computer was a long way off) in Art. 418 (pg. 432, [Fourier, 1878]) as

$$f(x) = \frac{1}{2\pi} \sum \int_a^b dx\, f(x) \cos \frac{2j\pi}{X} (x-a) \qquad (7.5)$$

which is rather hard to relate to, so in modern notation [Dominguez, 2016] the transform $Tf(u)$ from Equation 7.5 is calculated as

$$Tf(u) = \frac{2}{\pi} \int_0^{\infty} f(x) \cos(ux)\, dx \qquad (7.6)$$

FIGURE 7.1 Spectrum analysis by the alternative Fourier transform: (a) magnitude spectrum by alternative Fourier; (b) difference in magnitude spectrum; (c) difference in phase spectrum

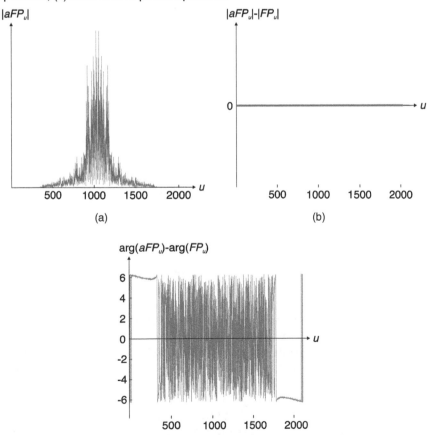

FIGURE 7.2   Fourier's notion of a transform [Fourier, 1878]

> 415. It is necessary to examine carefully the nature of the general propositions which serve to transform arbitrary functions: for the use of these theorems is very extensive, and we derive from them directly the solution of several important physical problems, which could be treated by no other method. The

So Fourier's first transform appears to have been a version of the cosine transform! Unlike some of the other transforms it is not eponymous, like the sine transform. Perhaps this is why: it is the original Fourier transform. In fact, Fourier was indeed the first to use the word transform, as shown in his book, here part of Art. 415 (pg. 425, [Fourier, 1878]) is shown in Figure 7.2, and this includes the notions that:

No one ever really claimed the cosine transform like the others: step up Prof. Cosine?

1. it concerns arbitrary functions; and
2. no other method allows the analysis.

This was far sighted indeed, laying the foundations of the new approach.

So Fourier actually invented the concept of a transform, though it was the cosine transform not the Fourier transform. Clearly, the plot thickens. There was in fact much-spirited debate about Fourier's original work, and in the best academic traditions, it was rejected at first and then became famous later when people realised how important it actually was. Lead on Macduff (one of the greatest misquotes)...

7.1.2   On the development of the Fourier transform

Fourier's original contribution was to challenge the central notions of the theory of heat transmission around the beginning of the 19th Century. At the time, the main notion was that heat flows from hot to cold according to the temperature difference between the two. (Thus it is similar to Ohm's law and electrical current, and that was developed only slightly later than Fourier's work.) We shall draw on Fourier's book [Fourier, 1878], which summarised his work on the Theory of Heat, and the title page is shown in Figure 7.3. In Google Scholar, at the time of writing it had received only 930 citations (by way of contrast, the original FFT paper has nearly 16,000). Fourier examined natural phenomena, stating that 'this class of phenomena has not as yet been submitted to a general analysis.' And later that 'the differential equations

FIGURE 7.3 Fourier's title page

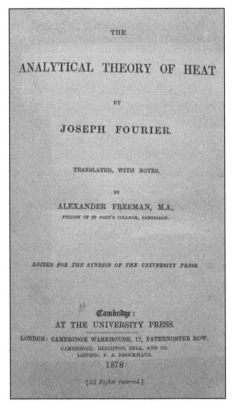

of the propagation of heat... reduce the physical questions to problems of pure analysis, and this is the proper object of theory.' One study has considered the basic nature of his claims and the differences of his new approach from those of his contemporaries and predecessors [Friedman, 1977]. Fourier's work was first published at the end of the year 1807, an extract from which was published in the *Bulletin des Sciences* (Société Philomatique, 1808, page 112). Fourier himself noted the difficulties in publication: 'The work we publish now has been written a long time since; different circumstances have delayed and often interrupted the printing of it' [Fourier, 1878]. This is a rather diplomatic way of saying that it was rejected, as it was by Lagrange.

The unease amongst Fourier's contemporaries centred on the notion that discontinuous functions could be represented by a sum of sinusoidal functions. This was seen early on in Figure 1.6 where the accuracy of reconstruction of a square wave depends on the number of Fourier descriptors used. Lagrange stated that it was impossible when Fourier presented his original work to the French Academy of Sciences, though they gave him a prize noting 'the novelty of the subject together with its importance' but adding 'while nevertheless observing that the manner in which the author arrives at his equations is not without difficulties'.

This might seem surprising to us now, but it shows how revolutionary Fourier's work was (one does wonder though if the objecting scientists had trouble with infinity too – even zero has had its problems in history). Sine waves and cosine waves are ubiquitous in our modern electronic society but that is now, not then. So publication was delayed until 1815 and followed by the original version of Fourier's book – in French – in 1822. The version we have used here is the English one, which was published in 1878. James Clerk Maxwell was not an original reviewer (he was born after it was originally rejected), though he was to write (in the English version) 'Fourier's treatise is one of the very few scientific books which can never be rendered antiquated by the progress of science.' High praise indeed.

Until computers arrived (and later the FFT) the Fourier transform evolved as a mathematical study, starting with the first book [Schlömilch, 1848] before even the English translation of Fourier's book had appeared. The term 'transform' started to emerge as a standard description [Titchmarsh, 1924], closely followed by a more definitive study [Wiener, 1933] (the same Wiener as in the Wiener–Khinchin theorem, Equation 2.43 [Titchmarsh, 1948]. Our modern computers were now starting to evolve, though the DFT on a processor with less power and storage than a modern vaper must have been a tedious task. Then we had the FFT (Section 3.6) and the rocket took off, fast. As previously mentioned, excellent coverage of the transform's history is available [Dominguez, 2016]. Our last task is to describe Joe himself.

7.2 BARON JEAN BAPTISTE JOSEPH FOURIER

Joseph Fourier was born in 1768 in Auxerre, France. He lived a tumultuous life in turbulent times often intersecting with Napoleon Bonaparte, as shown in his life's timeline in Figure 7.4. An image of Fourier, Figure 7.5(a) (courtesy of The Linda Hall Library of Science, Engineering & Technology) https://www.lindahall.org/joseph-fourier/ is not very flattering (an early Covid hairstyle?), and the earlier caricatures were derived from a portrait when he was younger. He does appear rather differently in all images so one wonders what he actually looked like. Fourier certainly had an interesting life: he was orphaned early and his interests developed most at a Benedictine Abbey and later at the Royal Military School of Auxerre. France was in revolutionary foment at the time and he joined Auxerre's revolutionary activities (anything

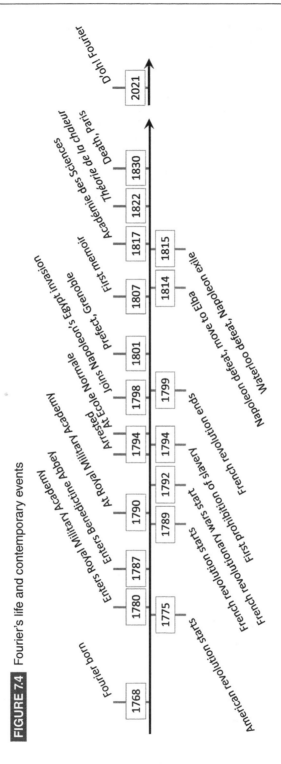

FIGURE 7.4 Fourier's life and contemporary events

FIGURE 7.5 Jean Baptiste Joseph Fourier, then and now: (a) Fourier [Reybaud, 1836]; (b) Fourier's grave; (c) Eiffel tower

(a)   (b)   (c)

was better than those indiscriminately used guillotines, though he was imprisoned). He was later proposed to the Ecole Normale in Paris just after it was founded to be taught by Lagrange and Laplace who were later to play an unattractive part in the Fourier transform's history. He joined Napoleon Bonaparte's expedition to Egypt as a scientific advisor where the activities of the British no doubt focussed his mind on improving the French cannonade. After the French were marooned in Egypt and being captured by the British forces, he later moved to Grenoble on Bonaparte's instigation 'the Prefect of the Department of Isère having recently died, I would like to express my confidence in citizen Fourier by appointing him to this place' where he continued his work on heat. The Joseph Fourier University, now part of Université Grenoble Alpes, was founded in 1811. His first work was presented in 1807, and rejected, though he later became Permanent Secretary of the French Academy of Sciences just after Bonaparte's defeat at Waterloo so he was clearly not without his supporters. Boney (as the Brits disparagingly called Napoleon) honoured him as a Baron in 1809, though Fourier aimed to avoid his march through Grenoble in 1814 by slipping out of the back door whilst Boney came through the front one. It appears to have been handbags at dawn since Fourier was awarded a pension which he never received, after Napoleon lost at Waterloo/Mont Saint-Jean/Belle Alliance (the protagonists have different names for the battle). Fourier's arterial walls had started to let him down in Egypt and he died of an aneurysm in 1830 and is buried at the Père Lachaise Cemetery in Paris, Figure 7.5(b). He joined the French scientists honoured on the Eifell Tower, Figure 7.5(c), and is included in Hawking's selection of the greatest mathematical works in history [Hawking, 2007]. He is also credited with the first observation of the greenhouse effect (it's in the Preliminary Discourse [Fourier, 1878]) though his eponymous transform is clearly his greatest legacy.

7.3 FINAL SUMMARY

Well, we hope you have enjoyed this book. There is a lot more that could be covered as this book allows you a glimpse into a fascinating subject. It is a focussed introduction aimed at a wide readership, with some jokes and diversions around the maths. Along with many scientists, Joseph Fourier certainly deserves greater recognition than is often accorded by modern society. Who cares about influencers and celebrities, when compared with a work that will last as long as our society, and which helps our lives in so many ways. Fourier's approach was new and contentious when he developed it, and thank crikey he had the tenacity to endure. Some of the material about its derivation, deployment and capabilities can be rather muddy waters in the modern literature. We have aimed to clarify this throughout, and there are many references which you can use for further study of this fascinating subject and they come after the Chapter 8, which are next and which summarise some of the material herein for ready reference. *Salut, et au revoir.*

Ready Reference Time

8.1 SUMMARY OF FOURIER TRANSFORMS AND THIER VARIANTS

Here we list all the transforms that are covered in this book and their principal advantages over and differences from the basic Fourier transforms (continuous and discrete).

TABLE 8.1 Summary of variants of the Fourier transform.

Transform	Section	Basis and differences/advantages over Fourier transform
Continuous Fourier	2.1	-
Cosine	5.1.1	Real transform, using cos without complex arithmetic
Discrete cosine (DCT)	5.1.2	Real transform, using cos without complex arithmetic
Discrete Fourier (DFT)	3.2	-
Discrete sine (DST)	5.1.2	… and using sin
Fast Fourier	3.6	Computational rearrangement of DFT for speed
Fastest Fourier in the East (FFTE)	3.6.6	Fastest FFT approach
Fastest Fourier in the West (FFTW)	3.6.6	Alternative to FFTE
Fourier–Mellin	5.5.2	Gives time/position and scale-invariant operation
Gabor wavelet	5.7.2	Gives optimal sensitivity to time and space
Hartley	5.3	Real transform, using cos and sin without complex arithmetic
Laplace	5.5.1	Differential operator for continuous systems analysis

(Continued)

TABLE 8.1 (Continued)

Transform	Section	Basis and differences/advantages over Fourier transform
Mellin	5.5.1	Gives scale-invariant operation
Morlet	5.7.2	Form of Gabor wavelet used in audio/seismic analysis
s	5.5.1	As Laplace
Short-Time Fourier (STFT)	2.5.1	Fourier transform of time section of signal selected by windowing function
Sine	5.1.1	Using sin as a basis
Radix 2 FFT	3.6.3	Computational arrangement of FFT
Walsh	5.2.1	Employs binary basis functions
Walsh-Hadamard	5.2.2	Employs binary basis functions effectively
Wavelet	5.7.2	Gives sensitivity to time and space
Winograd	3.6.6	Computational arrangement of FFT for speed
z	5.6	Unit delay operator for discrete systems analysis

8.2 SUMMARY OF PROPERTIES OF THE CONTINUOUS FOURIER TRANSFORM

TABLE 8.2 Summary of continuous Fourier transform properties.

Property	Time domain	Frequency domain
Superposition/ linearity	$p(t) + q(t)$	$\mathcal{F}(p(t) + q(t))$ $= \mathcal{F}(p(t)) + \mathcal{F}(q(t))$
Time shift	$\mathcal{F}(p(t - \tau))$	$Fp(\omega)\,e^{-j\omega\tau}$
Scaling in time	$p(\lambda t)$	$\dfrac{1}{\lambda}Fp\left(\dfrac{\omega}{\lambda}\right)$
Parseval's theorem (energy)	$\displaystyle\int_{-\infty}^{\infty}|p(t)|^2\,dt$	$\dfrac{1}{2\pi}\displaystyle\int_{-\infty}^{\infty}|Fp(\omega)|^2\,d\omega$
Differentiation	$\dfrac{dp(t)}{dt} = p'(t)$	$j\omega Fp(\omega)$
Modulation	$p(t) \times \cos(\omega_0 t)$	$\dfrac{1}{2}Fp(\omega - \omega_0) + \dfrac{1}{2}Fp(\omega + \omega_0)$

TABLE 8.2 (*Continued*)

Property	Time domain	Frequency domain
Convolution	$p(t) * q(t)$ $$= \int_{-\infty}^{\infty} p(\tau) q(t-\tau) \, d\tau$$	$\mathcal{F}(q(t)) \times \mathcal{F}(p(t))$
Correlation	$p(t) \otimes q(t)$ $$= \int_{-\infty}^{\infty} p(\tau) q(t+\tau) \, d\tau$$	$\mathcal{F}(q(-t)) \times \mathcal{F}(p(t))$
Autocorrelation	$p(t) \otimes p(t)$ $$= \int_{-\infty}^{\infty} p(\tau) p(t+\tau) \, d\tau$$	$\mathcal{F}(p(-t)) \times \mathcal{F}(p(t))$

Other properties		
Symmetry	$\|Fp(\omega)\|$ $= \|Fp(-\omega)\|$	$\arg(Fp(\omega))$ $= -\arg(Fp(-\omega))$ $Fp(\omega) = \overline{Fp(-\omega)}$
Uncertainty	variance $\left(\mathcal{F}\left(f(t)\right)\right) \times$ variance $\left(f(t)\right) = 1$	

8.3 CONTINUOUS FOURIER TRANSFORM PAIRS

We shall show the magnitudes of the transforms only notwithstanding the importance of phase.

TABLE 8.3 Continuous Fourier transform pairs.

Function	Time	Frequency
Sine wave	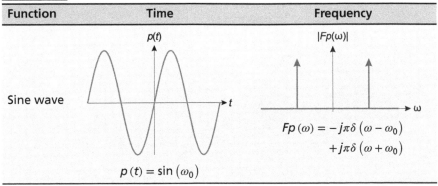 $p(t) = \sin(\omega_0)$	$Fp(\omega) = -j\pi\delta(\omega - \omega_0)$ $+ j\pi\delta(\omega + \omega_0)$

(*Continued*)

TABLE 8.3 (Continued)

Function	Time	Frequency

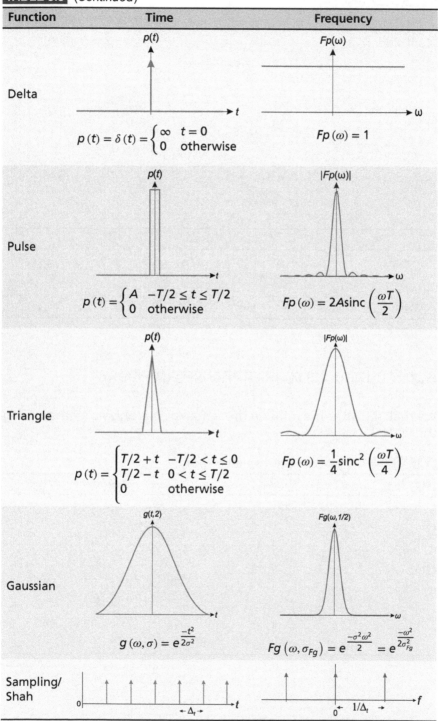

Delta

$$p(t) = \delta(t) = \begin{cases} \infty & t = 0 \\ 0 & \text{otherwise} \end{cases}$$

$$Fp(\omega) = 1$$

Pulse

$$p(t) = \begin{cases} A & -T/2 \le t \le T/2 \\ 0 & \text{otherwise} \end{cases}$$

$$Fp(\omega) = 2A\text{sinc}\left(\frac{\omega T}{2}\right)$$

Triangle

$$p(t) = \begin{cases} T/2 + t & -T/2 < t \le 0 \\ T/2 - t & 0 < t \le T/2 \\ 0 & \text{otherwise} \end{cases}$$

$$Fp(\omega) = \frac{1}{4}\text{sinc}^2\left(\frac{\omega T}{4}\right)$$

Gaussian

$$g(\omega, \sigma) = e^{\frac{-t^2}{2\sigma^2}}$$

$$Fg\left(\omega, \sigma_{Fg}\right) = e^{\frac{-\sigma^2\omega^2}{2}} = e^{\frac{-\omega^2}{2\sigma_{Fg}^2}}$$

Sampling/Shah

8.4 SUMMARY OF PROPERTIES OF THE DISCRETE FOURIER TRANSFORM

TABLE 8.4 Summary of discrete Fourier transform properties.

Property	Time domain	Frequency domain
Superposition/ linearity	$\mathbf{p} + \mathbf{q}$	$\mathcal{F}(\mathbf{p} + \mathbf{q})$ $= \mathcal{F}(\mathbf{p}) + \mathcal{F}(\mathbf{q})$
Time shift	$\mathcal{F}(\mathbf{p}[x - \Delta])$	$\mathbf{Fp}[u] \times e^{-j\omega\Delta}$
Scaling in time	$\mathbf{p}[x, k] = \begin{cases} \mathbf{p}[x/k] & n = km \\ 0 & otherwise \end{cases}$	$\mathbf{Fp}[ku]$
Parseval's theorem (energy)	$\sum_{x=0}^{N-1} \lvert(p_x)^2\rvert$	$\frac{1}{N} \sum_{u=0}^{N-1} \lvert(Fp_u)^2\rvert$
Symmetry		$\lvert Fp_{(N-1)-u}\rvert = \lvert Fp_u\rvert$
Differentiation	$\mathbf{p}' \equiv \mathbf{p}[m] - \mathbf{p}[m-1]$	$(1 - e^{-j\omega})\,\mathcal{F}(\mathbf{p})$
Convolution	$\mathbf{p} * \mathbf{q} = \mathbf{p}[m] * \mathbf{q}[m]$	$\mathcal{F}(\mathbf{p}).\times \mathcal{F}(\mathbf{q})$
Correlation	$\mathbf{p} \otimes \mathbf{q} = \mathbf{p}[m] \otimes \mathbf{q}[m]$	$\mathcal{F}(\mathbf{p}).\times \overline{\mathcal{F}(\mathbf{q})}$
Autocorrelation	$\mathbf{p} \otimes \mathbf{p} = \mathbf{p}[m] \otimes \mathbf{p}[m]$	$\mathcal{F}(\mathbf{p}).\times \overline{\mathcal{F}(\mathbf{p})}$

8.5 DISCRETE FOURIER TRANSFORM PAIRS

As with the continuous versions, we shall show the magnitudes of the transforms only notwithstanding the importance of phase. The signals and transforms are displayed as centred (d.c. = N/2 for N points), whereas the maths is not always centred (d.c. = 0), noting shift invariance.

TABLE 8.5 Discrete Fourier transform pairs.

Function	Time	Frequency
Sine wave	$p_x = \sin\left(\dfrac{2\pi f x}{N}\right)$	$Fp_u = \begin{cases} 0.5 & u = N/2 \pm f \\ 0 & otherwise \end{cases}$

(Continued)

TABLE 8.5 (*Continued*)

Function	Time	Frequency

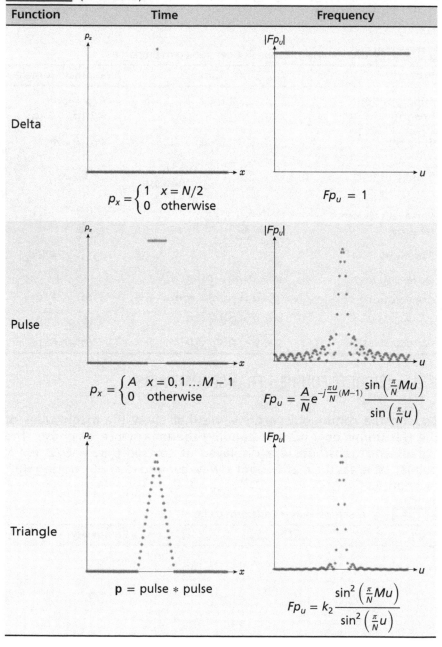

Delta

$$p_x = \begin{cases} 1 & x = N/2 \\ 0 & \text{otherwise} \end{cases}$$

$$Fp_u = 1$$

Pulse

$$p_x = \begin{cases} A & x = 0, 1 \dots M - 1 \\ 0 & \text{otherwise} \end{cases}$$

$$Fp_u = \frac{A}{N} e^{-j\frac{\pi u}{N}(M-1)} \frac{\sin\left(\frac{\pi}{N}Mu\right)}{\sin\left(\frac{\pi}{N}u\right)}$$

Triangle

$$\mathbf{p} = \text{pulse} * \text{pulse}$$

$$Fp_u = k_2 \frac{\sin^2\left(\frac{\pi}{N}Mu\right)}{\sin^2\left(\frac{\pi}{N}u\right)}$$

TABLE 8.5 (*Continued*)

Function	Time	Frequency
Gaussian	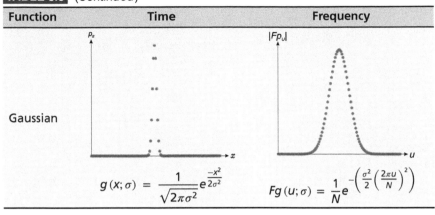	

$$g(x;\sigma) = \frac{1}{\sqrt{2\pi\sigma^2}}e^{\frac{-x^2}{2\sigma^2}}$$

$$Fg(u;\sigma) = \frac{1}{N}e^{-\left(\frac{\sigma^2}{2}\left(\frac{2\pi u}{N}\right)^2\right)}$$

References

Agaian, S.S., Panetta, K. and Grigoryan, A.M. 2001. Transform-based image enhancement algorithms with performance measure. *IEEE Transactions on Image Processing, 10*(3), pp.367–382.

Aguado, A.S., Montiel, E. and Nixon, M.S. 2000. On the intimate relationship between the principle of duality and the Hough transform. *Proceedings of the Royal Society of London. Series A: Mathematical, Physical and Engineering Sciences, 456*(1995), pp.503–526.

Ahmed, N., Natarajan, T. and Rao, K.R. 1974. Discrete cosine transform. *IEEE Transactions on Computers, 100*(1), pp.90–93.

Ahmed, N., Rao, K.R. and Abdussattar, A.L. 1973. On cyclic autocorrelation and the Walsh-Hadamard transform. *IEEE Transactions on Electromagnetic Compatibility,* (3), pp.141–146.

Alonso-Fernandez, F., Fierrez, J., Ortega-Garcia, J., Gonzalez-Rodriguez, J., Fronthaler, H., Kollreider, K. and Bigun, J. 2007. A comparative study of fingerprint image-quality estimation methods. *IEEE Transactions on Information Forensics and Security, 2*(4), pp.734–743.

Alsteris, L.D. and Paliwal, K.K. 2007. Short-time phase spectrum in speech processing: A review and some experimental results. *Digital Signal Processing, 17*(3), pp.578–616.

Altman, J., Reitbock, H.J.P. 1984. A fast correlation method for scale- and translation-invariant pattern recognition. *IEEE Transactions on PAMI, 6*(1), pp.46–57.

Ayinala, M. and Parhi, K.K. 2013. FFT architectures for real-valued signals based on radix-2^3 and radix-2^4 algorithms. *IEEE Transactions on Circuits and Systems I, 60*(9), pp.2422–2430.

Banham, M.R. and Katsaggelos, A.K. 1996. Spatially adaptive wavelet-based multiscale image restoration. *IEEE Transactions on Image Processing, 5*(4), pp.619–634.

Beauchamp, K.G. 1975. *Walsh Functions and Their Applications.* Academic Press Inc.

Benesty, J., Cohen, I. and Chen, J. 2018. *Fundamentals of Signal Enhancement and Array Signal Processing.* John Wiley & Sons Singapore Pvt. Limited.

Bertrand, J., Bertrand, P., and Ovarlez, J.P. 2000. The Mellin transform. In *The Transforms and Applications Handbook,* 2nd Ed. Boca Raton: CRC Press LLC.

Bovik, A.C., Clark, M. and Geisler, W.S. 1990. Multichannel texture analysis using localized spatial filters. *IEEE Transactions on PAMI, 12*(1), pp.55–73.

Bracewell, R.N. 1983. Discrete hartley transform. *JOSA, 73*(12), pp.1832–1835.

Bracewell, R.N. 1984. The fast Hartley transform. *Proceedings of the IEEE, 72*(8), pp.1010–1018.

Bracewell, R.N. 1989. The Fourier transform. *Scientific American, 260*(6), pp.86–95.

Bracewell, R.N. 1986. *The Fourier Transform and Its Applications,* Revised 2nd Ed. McGraw-Hill, Book Co., Singapore.

Britanak, V., Yip, P.C. and Rao, K.R. 2010. *Discrete Cosine and Sine Transforms: General Properties, Fast Algorithms and Integer Approximations.* Elsevier.

Brodatz, P. 1966. *Textures: A Photographic Album for Artists and Designers.* Dover Pubns.

Broughton, S. and Bryan, K. 2018. *Discrete Fourier Analysis and Wavelets: Application to Signal and Image Processing,* 2nd Ed. Wiley.

Brown, J.I. 2004. Mathematics, physics and a hard day's night. *CMS Notes, 36*(6), pp.4–8.

Buades, A., Coll, B. and Morel, J.M. 2005. A non-local algorithm for image denoising. In *2005 IEEE CVPR* (Vol. 2, pp.60–65).

Bull, D.R. 2014. *Communicating Pictures: A Course in Image and Video Coding.* Academic Press.

Campisi, P. and Egiazarian, K. eds. 2017. *Blind Image Deconvolution: Theory and Applications.* CRC press.

Casasent, D. and Psaltis, D. 1977. New optical transforms for pattern recognition. *Proceedings of the IEEE, 65*(1), pp.77–83.

Cheng, C. and Yu, F. 2015. An optimum architecture for continuous-flow parallel bit reversal. *IEEE Signal Processing Letters, 22*(12), pp.2334–2338.

Cipolla, R., Battiato, S. and Farinella, G.M. eds. 2014. *Registration and Recognition in Images and Videos.* Springer.

Coifman, R.R. and Donoho, D.L. 1995. Translation-invariant de-noising. In *Wavelets and Statistics* (pp.125–150). Springer, New York, NY.

Cooley, J.W., Lewis, P.A. and Welch, P.D. 1967. Historical notes on the fast Fourier transform. *Proceedings of the IEEE, 55*(10), pp.1675–1677.

Cooley, J.W. and Tukey, J.W. 1965. An algorithm for the machine calculation of complex Fourier series. *Mathematics of Computation, 19*(90), pp.297–301.

Corinthios, M. 2009. *Signals, systems, transforms, and digital signal processing with MATLAB.* CRC Press.

Cunado, D., Nixon, M.S. and Carter, J.N. 1997. Using gait as a biometric, via phase-weighted magnitude spectra. In *International Conference on Audio- and Video-Based Biometric Person Authentication* (pp.93–102). Springer, Berlin, Heidelberg.

Cunado, D., Nixon, M.S. and Carter, J.N. 2003. Automatic extraction and description of human gait models for recognition purposes. *Computer Vision and Image Understanding, 90*(1), pp.1–41.

da Silva, E.A. and Ghanbari, M. 1996. On the performance of linear phase wavelet transforms in low bit-rate image coding. *IEEE Transactions on Image Processing, 5*(5), pp.689-704.

Dalí. https://www.dalipaintings.com/

Daubechies, I. 2009. *The wavelet transform, time-frequency localization and signal analysis* (pp.442–486). Princeton University Press.

Daugman, J.G. 1988. Complete discrete 2-D Gabor transforms by neural networks for image analysis and compression. *IEEE Transactions on Acoustics, Speech, and Signal Processing, 36*(7), pp.1169–1179.

Daugman, J.G. 1993. High confidence visual recognition of persons by a test of statistical independence. *IEEE Transactions on PAMI, 15*(11), pp.1148–1161.

De Castro, E. and Morandi, C. 1987. Registration of translated and rotated images using finite Fourier transforms. *IEEE Transactions on PAMI, 5*, pp.700–703.

Deng, G. 2010. A generalized unsharp masking algorithm. *IEEE Transactions on Image Processing, 20*(5), pp.1249–1261.

Dominguez, A. 2016. Highlights in the history of the Fourier transform. *IEEE Pulse, 7*(1), pp.55–61.

Donoho, D.L. 2016. Compressed sensing. *IEEE Transactions on Information Theory, 52*(4), pp.1289–1306.

Fourier, J.B.J., 1878. *The Analytical Theory of Heat.* The University Press.

Friedman, R.M. 1977. The creation of a new science: Joseph Fourier's analytical theory of heat. *Historical Studies in the Physical Sciences, 8*, pp.73–99.

Frigo, M. and Johnson, S.G. 2005. The design and implementation of FFTW3. *Proceedings of the IEEE, 93*(2), pp.216–231.

Garrido, M., Grajal, J. and Gustafsson, O. 2011. Optimum circuits for bit reversal. *IEEE Transactions on Circuits and Systems II: Express Briefs, 58*(10), pp.657–661.

Goodman, J.W. 2005. *Introduction to Fourier Optics.* Roberts and Company Publishers.

Goupillaud, P., Grossmann, A. and Morlet, J. 1984. Cycle-octave and related transforms in seismic signal analysis. *Geoexploration, 23*(1), pp.85–102.

Graves, A., Mohamed, A.R. and Hinton, G. 2013. Speech recognition with deep recurrent neural networks. In *2013 IEEE International Conference on Acoustics, Speech and Signal Processing* (pp.6645–6649).

Gu, J., Wang, Z., Kuen, J., Ma, L., Shahroudy, A., Shuai, B., Liu, T., Wang, X., Wang, G., Cai, J. and Chen, T. 2018. Recent advances in convolutional neural networks. *Pattern Recognition, 77*, pp.354–377.

Gupta, A. and Kumar, V. 1993. The scalability of FFT on parallel computers. *IEEE Transactions on Parallel and Distributed Systems, 4*(8), pp.922–932.

Hahn, B. and Valentine, D. 2016. *Essential MATLAB for Engineers and Scientists,* Academic Press, 7th Ed.

Hamood, M.T. and Boussakta, S. 2011. Fast Walsh–Hadamard–Fourier transform algorithm. *IEEE Transactions on Signal Processing, 59*(11), pp.5627–5631.

Harris, F.J. 1978. On the use of windows for harmonic analysis with the discrete Fourier transform. *Proceedings of the IEEE, 66*(1), pp.51–83.

Hartley, R.V. 1942. A more symmetrical Fourier analysis applied to transmission problems. *Proceedings of the IRE, 30*(3), pp.144–150.

Hawking, S. ed. 2007. *God Created the Integers: The Mathematical Breakthroughs That Changed History.* Running Press Adult.

Heinzel, G., Rüdiger, A. and Schilling, R. 2002. Spectrum and spectral density estimation by the Discrete Fourier transform (DFT), including a comprehensive list of window functions and some new at-top windows. Report, Max-Planck-Institut fur Gravitationsphysik, Hannover, Germany.

Hong, L., Wan, Y. and Jain, A. 1998. Fingerprint image enhancement: Algorithm and performance evaluation. *IEEE Transactions on PAMI, 20*(8), pp.777–789.

Hurley, D.J., Nixon, M.S. and Carter, J.N. 2002. Force field energy functionals for image feature extraction. *Image and Vision computing, 20*(5-6), pp.311–317.

Hurley, D.J., Nixon, M.S. and Carter, J.N. 2005. Force field feature extraction for ear biometrics. *Computer Vision and Image Understanding, 98*(3), pp.491–512.

Ifeachor, E.C. and Jervis, B.W. 2002. *Digital Signal Processing: A Practical Approach.* Pearson Education.

Jain, A.K., Prabhakar, S., Hong, L. and Pankanti, S. 2000. Filterbank-based fingerprint matching. *IEEE Transactions on Image Processing, 9*(5), pp.846–859.

Johnson, S.G. and Frigo, M. 2006. A modified split-radix FFT with fewer arithmetic operations. *IEEE Transactions on Signal Processing, 55*(1), pp.111–119.

Juang, B.H. and Rabiner, L.R. 1991. Hidden Markov models for speech recognition. *Technometrics, 33*(3), pp.251–272.

Karp, A.H. 1996. Bit reversal on uniprocessors. *SIAM Review, 38*(1), pp.1–26.

Katsaggelos, A.K. 2012. *Digital Image Restoration.* Springer Publishing Company, Incorporated.

Klein, M.V. and Furtak, T.E. 1986. *Optics.* John Wiley and Sons, New York, NY.

Kolba, D. and Parks, T.W. 1977. A prime factor FFT algorithm using high-speed convolution. *IEEE Transactions on Acoustics, Speech, and Signal Processing, 25*(4), pp.281–294.

Kovesi, P. 1999. Image features from phase congruency. *Videre: Journal of Computer Vision Research, 1*(3), pp.1–26. http://citeseerx.ist.psu.edu/viewdoc/summary?doi=10.1.1.54.5658

Kundur, D. and Hatzinakos, D. 1996. Blind image deconvolution revisited. *IEEE Signal Processing Magazine, 13*(6), pp.61–63.

Lades, M., Vorbruggen, J.C., Buhmann, J., Lange, J., Von Der Malsburg, C., Wurtz, R.P. and Konen, W. 1993. Distortion invariant object recognition in the dynamic link architecture. *IEEE Transactions on Computers, 42*(3), pp.300–311.

Laine, A. and Fan, J. 1993. Texture classification by wavelet packet signatures. *IEEE Transactions on PAMI, 15*(11), pp.1186–1191.

Lavin, A. and Gray, S. 2016. Fast algorithms for convolutional neural networks. In *Proceedings of the IEEE Conference on Computer Vision and Pattern Recognition* (pp.4013–4021).

Lex, T. C. O. L. T. 2012. *Who is Fourier, a Mathematical Adventure,* 2nd Ed. Language Research Foundation, Boston, MA USA.

Liu, L., Chen, J., Fieguth, P., Zhao, G., Chellappa, R. and Pietikäinen, M. 2019. From BoW to CNN: Two decades of texture representation for texture classification. *International Journal of Computer Vision, 127*(1), pp.74–109.

Liu, L., Fieguth, P., Guo, Y., Wang, X. and Pietikäinen, M. 2017. Local binary features for texture classification: Taxonomy and experimental study. *Pattern Recognition, 62,* pp.135–160.

Liu, S.S. and Jernigan, M.E. 1990. Texture analysis and discrimination in additive noise. *Computer Vision, Graphics, and Image Processing, 49*(1), pp.52–67.

Ma, Z.G., Yin, X.B. and Yu, F. 2015. A novel memory-based FFT architecture for real-valued signals based on a radix-2 decimation-in-frequency algorithm. *IEEE Transactions on Circuits and Systems II: Express Briefs, 62*(9), pp.876–880.

Maintz, J.A. and Viergever, M.A. 1998. A survey of medical image registration. *Medical Image Analysis, 2*(1), pp.1–36.

Maio, D. and Maltoni, D. 1997. Direct gray-scale minutiae detection in finger-prints. *IEEE Transactions on PAMI, 19*(1), pp.27–40.

Mallat, S. 2008. *A Wavelet Tour of Signal Processing.* Elsevier.

Martucci, S.A. 1994. Symmetric convolution and the discrete sine and cosine transforms. *IEEE Transactions on Signal Processing, 42*(5), pp.1038–1051.

Morrone, M.C. and Owens, R.A. 1987. Feature detection from local energy. *Pattern Recognition Letters, 6*(5), pp.303–313.

Mulet-Parada, M. and Noble, J.A. 2000. 2D+T acoustic boundary detection in echocardiography. *Medical Image Analysis, 4*(1), pp.21–30.

Myerscough, P.J. and Nixon, M.S. 2004. Temporal phase congruency. In *6th IEEE Southwest Symposium on Image Analysis and Interpretation* (pp.76–79).

Ngan, K.N., Yap, C.W. and Tan, K.T. 2018. *Video Coding for Wireless Communication Systems* (Vol. 2). CRC Press.

Nikolić, M., Jović, A.,Jakić, J., Slavnić, V. and Balaž, A. 2014. An analysis of FFTW and FFTE performance. In *High-Performance Computing Infrastructure for Southeast Europe's Research Communities* (pp.163–170). Springer, Cham.

Nixon, M. 2015. *Digital Electronics: A Primer: Introductory Logic Circuit Design* (Vol. 1). World Scientific Publishing Company.

Nixon, M. and Aguado, A. 2019. *Feature Extraction and Image Processing for Computer Vision,* 4th Ed. Academic Press.

Nixon, M.S., Tan, T. and Chellappa, R. 2010. *Human Identification Based on Gait.* Springer Science & Business Media.

OpenCV Template Operator. https://github.com/opencv/opencv/tree/master/modules/imgproc/src

Oppenheim, A.V. and Lim, J.S. 1981. The importance of phase in signals. *Proceedings of the IEEE, 69*(5), pp.529–541.

Oppenheim, A.V., Schafer, R.W. and Buck, J.R. 1999. *Discrete-Time Signal Processing,* 2nd Ed. Prentice Hall, Upper Saddle River.

Oppenheim, A.V., Willsky, A.S. and Hamid, S. 1997. *Signals and Systems,* Processing series. Prentice Hall, Hoboken, NJ.

Ortega, A., Frossard, P., Kovačević, J., Moura, J.M. and Vandergheynst, P. 2018. Graph signal processing: Overview, challenges, and applications. *Proceedings of the IEEE, 106*(5), pp.808–828.

Osgood, B.G., 2019. *Lectures on the Fourier Transform and Its Applications* (Vol. 33). American Mathematical Society.

Pan, W., Qin, K. and Chen, Y. 2008. An adaptable-multilayer fractional Fourier transform approach for image registration. *IEEE Transactions on PAMI, 31*(3), pp.400–414.

Petrou, M. and Sevilla, P.G., 2006. *Image Processing: Dealing with Texture.* Wiley, New York, NY.

Polesel, A., Ramponi, G. and Mathews, V.J. 2000. Image enhancement via adaptive unsharp masking. *IEEE Transactions on Image Processing, 9*(3), pp.505–510.

Povey, D., Ghoshal, A., Boulianne, G., Burget, L., Glembek, O., Goel, N., Hannemann, M., Motlicek, P., Qian, Y., Schwarz, P. and Silovsky, J. 2011. The Kaldi speech recognition toolkit. In *IEEE 2011 Workshop on Automatic Speech Recognition and Understanding.*

Pratt, W.K. 1992. *Digital Image Processing.* Wiley.

Pratt, W.K., Kane, J. and Andrews, H.C. 1969. Hadamard transform image coding. *Proceedings of the IEEE, 57*(1), pp.58–68.

Randen, T. and Husoy, J.H. 1999. Filtering for texture classification: A comparative study. *IEEE Transactions on PAMI, 21*(4), pp.291–310.

Rao, K.R. and Yip, P. 2014. *Discrete Cosine Transform: Algorithms, Advantages, Applications,* Academic Press.

Reddy, B.S. and Chatterji, B.N. 1996. An FFT-based technique for translation, rotation, and scale-invariant image registration. *IEEE Transactions on Image Processing, 5*(8), pp.1266–1271.

Reybaud, L. 1836. *Histoire de l'expédition Française en Égypte* (Vol. 8, opp. page 36). A.-J. Dénain.

Robinson, G. 1972. Logical convolution and discrete Walsh and Fourier power spectra. *IEEE Transactions on Audio and Electroacoustics, 20*(4), pp.271–280.

Salehi, S.A., Amirfattahi, R. and Parhi, K.K. 2013. Pipelined architectures for real-valued FFT and Hermitian-symmetric IFFT with real datapaths. *IEEE Transactions on Circuits and Systems II: Express Briefs, 60*(8), pp.507–511.

Schlömilch,O.X. 1848. *Analytische studien: die Fourier'schen reihen und integrale* (Vol. 2). Verlag von Wilhem Engelmann, Leipzig.

Schulz, T.J., Stribling, B.E. and Miller, J.J. 1997. Multiframe blind deconvolution with real data: Imagery of the Hubble Space Telescope. *Optics Express, 1*(11), pp.355–362.

Seely, R.D., Tao, M.W., Lindskog, A. and Anneheim, G.T., Apple Inc, 2019. *Shallow Depth of Field Rendering.* U.S. Patent 10,410,327.

Selesnick, I.W., Baraniuk, R.G. and Kingsbury, N.C. 2005. The dual-tree complex wavelet transform. *IEEE Signal Processing Magazine, 22*(6), pp.123–151.

Shanks, J.L. 1969. Computation of the fast Walsh-Fourier transform. *IEEE Transactions on Computers, 100*(5), pp.457–459.

Smith III, J.O. 2011. *Spectral Audio Signal Processing.* W3K Publishing. https://ccrma.stanford.edu/~jos/sasp/sasp.html

Spanias, A.S. 1994. Speech coding: A tutorial review. *Proceedings of the IEEE, 82*(10), pp.1541–1582.

Strum, R.D. and Krik, D.E. 1988. *First principles of discrete systems and digital signal processing.* Addison-Wesley Longman Publishing Co., Inc.

Talwalkar, S.A. and Marple, S.L. 2010. Time-frequency scaling property of discrete Fourier transform (DFT). In *2010 IEEE International Conference on Acoustics, Speech and Signal Processing* (pp.3658–3661).

Tanaka, Y., Eldar, Y.C., Ortega, A. and Cheung, G. 2020. Sampling signals on graphs: From theory to applications. *IEEE Signal Processing Magazine, 37*(6), pp.14–30.

Thompson, D.F. ed. 1996. *The Oxford Modern English Dictionary.* Oxford University Press, USA.

Titchmarsh, E.C. 1924. Hankel transforms. In *Mathematical Proceedings of the Cambridge Philosophical Society* (Vol. 22, No. 1, pp.67–67). Cambridge University Press.

Titchmarsh, E.C. 1948. *Introduction to the Theory of Fourier Integrals,* 2nd Ed. Clarendon Press.

Tun, I.D. and Lee, S.U. 1993. On the fixed-point-error analysis of several fast DCT algorithms. *IEEE Transactions on Circuits and Systems for Video Technology, 3*(1), pp.27–41.

Van Veen, B.D. and Buckley, K.M. 1988. Beamforming: A versatile approach to spatial filtering. *IEEE ASSP magazine, 5*(2), pp.4–24.

Viergever, M.A., Maintz, J.A., Klein, S., Murphy, K., Staring, M. and Pluim, J.P. 2016. A survey of medical image registration–under review. *Medical Image Analysis, 33*, pp.140–144.

Vonnegut, K. 1973. *Breakfast of Champions, or Goodbye Blue Monday.* Delacourte Press.

Walsh, J. L. 1923. A closed set of normal orthogonal functions. *American Journal of Mathematics, 45*(1), pp.5–24.

Weilong, C., Meng, J.E. and Shiqian, W. 2005. PCA and LDA in DCT domain. *Pattern Recognition Letters, 26*(15), pp.2474–2482.

Wiener, N. 1933. *The Fourier Integral and Certain of its Applications.* Cambridge University Press, Cambridge, UK.

Winograd, S. 1978. On computing the discrete Fourier transform. *Mathematics of Computation, 32*(141), pp.175–199.

Wu, G., Masia, B., Jarabo, A., Zhang, Y., Wang, L., Dai, Q., Chai, T. and Liu, Y. 2017. Light field image processing: An overview. *IEEE Journal of Selected Topics in Signal Processing, 11*(7), pp.926–954.

Xiang, Y., Wang, F., Wan, L. and You, H. 2017. SAR-PC: Edge detection in SAR images via an advanced phase congruency model. *Remote Sensing, 9*(3), p.209.

Zheng, Y., Nixon, M.S. and Allen, R. 2004. Automated segmentation of lumbar vertebrae in digital videofluoroscopic images. *IEEE Transactions on Medical Imaging*, 23(1), pp.45–52.

Zitova, B. and Flusser, J. 2003. Image registration methods: A survey. *Image and Vision Computing, 21*(11), pp.977–1000.

Zokai, S. and Wolberg G. 2005. Image registration using log-polar mappings for recovery of large-scale similarity and projective transformations. *IEEE Transactions on Image Processing, 14,* pp.1422–1434.

Index

Lightning Source UK Ltd.
Milton Keynes UK
UKHW021826250322
400637UK00005B/90